THE COMMONWEALTH AND INTERNATIONAL LIBRARY

Joint Chairmen of the   SIR ROBERT ROBINSON, O.M., F.R.S.,
Honorary Editorial Advisory Board   London
                              DEAN ATHELSTAN SPILHAUS,
                              Minnesota

Publisher   ROBERT MAXWELL, M.C., M.P.

SELECTED READINGS IN PHYSICS

General Editor   D. TER HAAR

VELOCITY OF LIGHT

VELOCITY OF LIGHT

J. H. SANDERS
Fellow of Oriel College, Oxford

PERGAMON PRESS
OXFORD · LONDON · EDINBURGH · NEW YORK
PARIS · FRANKFURT

Pergamon Press Ltd., Headington Hill Hall, Oxford
4 & 5 Fitzroy Square, London W.1

Pergamon Press (Scotland) Ltd., 2 & 3 Teviot Place, Edinburgh 1

Pergamon Press Inc., 122 East 55th Street, New York 10022

Pergamon Press GmbH, Kaiserstrasse 75, Frankfurt-am-Main

Copyright © 1965 Pergamon Press Ltd.
First Edition 1965
Library of Congress Catalog Card No. 65-19840

Set in 10 on 12 pt. Times
and Printed in Great Britain
by Bell and Bain Ltd., Glasgow

This book is sold subject to the condition
that it shall not, by way of trade, be lent,
resold, hired out, or otherwise disposed
of without the publisher's consent,
in any form of binding or cover
other than that in which
it is published.

(2338/65)

Contents

Preface ... vii

Introduction ... ix

Part 1

I	Early Work on the Velocity of Light	3
II	Michelson, Pease and Pearson	9
III	The Period 1935–46	16
IV	Essen and Gordon-Smith	19
V	Bergstrand	25
VI	More Recent Measurements	32
VII	Future Measurements of the Velocity of Light	36

References ... 39

Part 2

1	*Measurement of the velocity of light in a partial vacuum*, by A. A. Michelson, F. G. Pease and F. Pearson	43
2	*The velocity of propagation of electromagnetic waves derived from the resonant frequencies of a cylindrical cavity resonator*, by L. Essen and A. C. Gordon-Smith	80
3	*A determination of the velocity of light*, by E. Bergstrand	101
	Index	143

Preface

THE main part of this book is concerned with three papers which describe recent measurements of the velocity of light. The three papers are reprinted in full: two of them, at least, may not be readily accessible except in the larger scientific libraries. In Part 1 a commentary provides a background for the work described in the papers, and an explanation of some points which have necessarily been treated briefly in them. To place the papers in the right perspective a short survey is given of early work on the velocity of light, and the final sections describe the most recent measurements, and the pattern which future work may be expected to follow.

I am indebted to the University of Chicago Press for permission to reprint the paper by Michelson, Pease and Pearson which appeared in the *Astrophysical Journal*, to the Royal Society, Dr. L. Essen and Mr. A. C. Gordon-Smith for permission to reprint the paper from the *Proceedings of the Royal Society*, and to the Royal Swedish Academy of Science and Dr. Bergstrand for permission to reprint the paper from the *Arkiv för Fysik*.

J. H. SANDERS

The internationally recommended value of the velocity of light is at present

$$299{,}792 \cdot 5 \text{ km/sec};$$

the uncertainty of this value is about one part per million.

Introduction

THE precise measurement of fundamental physical quantities is a subject which has commanded the attention of some of the world's most notable experimental physicists. Of all the measurable constants of nature the velocity of light has the most immediate appeal: light is a directly observable phenomenon, and the philosopher is led immediately to speculate on its nature and how fast it travels. These problems can only be solved by experiment. Light has such a high velocity that, unlike the case of sound, there is no simple way of distinguishing whether it travels from one place to another instantaneously or within a definite time interval.

Astronomical data provided, towards the end of the seventeenth century and the beginning of the eighteenth, the first evidence for the finite value of the velocity of light. There was then a long interval, until the middle of the nineteenth century, before the development of techniques which initiated a series of laboratory measurements of the velocity. At about this time theoretical physics began to reveal that the velocity of light plays a much more fundamental part in physics than had hitherto been realized. Maxwell's electromagnetic theory, and the subsequent experimental developments in the production of electromagnetic radiation, showed that visible light is only part of a broad spectrum of radiation, all of which is propagated in free space with the same velocity. Moreover, the connection between electric and magnetic fields expressed in Maxwell's equations allows this velocity to be expressed in a fundamental way related to the units used to measure electric and magnetic quantities. Thus, in the c.g.s. system of units, the quantity c, which appears in the relations between electrostatic and electromagnetic units, is identical with the velocity of electromagnetic waves in free space; in the m.k.s. system of units the corresponding quantity is $(\varepsilon_0 \mu_0)^{-\frac{1}{2}}$, where ε_0 and μ_0 are the so-called permittivity and permeability of free space respectively.

The development of the special theory of relativity by Einstein showed that a quantity c, which again can be identified with the velocity of electromagnetic radiation in free space, plays a fundamental part in this branch of physics. It is the upper limit of the velocity of communication between two points in space, the limiting velocity of matter or energy relative to an observer, and appears as a constant connecting the units of mass and the units of energy.

The use of the term " the velocity of light " to mean the speed of electromagnetic radiation in free space (i.e. in an unbounded vacuum) has become so widespread that it is almost pedantic to depart from its use. Unless it is made clear that a more restricted meaning is intended, the shorter term (in spite of the objections that can be raised, for example, the use of the vector quantity "velocity " rather than " speed ") will be used throughout the commentary in this book.

Three outstanding papers in the recent history of the measurement of the velocity of light have been chosen for study: " Measurement of the velocity of light in a partial vacuum ", by A. A. Michelson, F. G. Pease and F. Pearson, published in Volume 82 of the *Astrophysical Journal*, 1935; " The velocity of propagation of electromagnetic waves derived from the resonant frequencies of a cylindrical cavity resonator ", by L. Essen and A. C. Gordon-Smith, published in the *Proceedings of the Royal Society*, Volume A 194, 1948; and " A determination of the velocity of light ", by E. Bergstrand, published (in English) in Volume 2 of the *Arkiv för Fysik*, 1950.

The first of these papers is an account of a well-known experiment which, we now know, produced an erroneous result. The second is the first full description of a determination which disagreed with Michelson, Pease and Pearson's value and gave a result in accord with the value now accepted. Bergstrand's paper describes a measurement modelled along classical lines but using modern electronic techniques; the quoted result is as precise as any now available and in complete agreement with recent independent measurements. The first of these papers may be regarded as both challenging and instructive, and the reader is invited to attempt to discover the source of the error in the quoted result.

Part 1

1
Early Work on the Velocity of Light

The first indication that light travels with a finite velocity was obtained by Römer in 1676. Römer discovered that the intervals between the eclipses of Jupiter's first satellite by the planet were smaller when the earth was approaching Jupiter than when the earth and Jupiter were receding from each other. He interpreted this as due to the time taken for light to travel through space; it is, in fact, an example of the Doppler effect. His value of 48,203 leagues per second (or about 214,000 km/sec) is of the correct order of magnitude, but too small because of his inaccurate knowledge of the relative velocity of the earth and Jupiter.

Römer's discovery that light is propagated with a finite velocity rather than instantaneously, a view held, for example, by both Kepler and Descartes, was not universally accepted until 1727 when Bradley demonstrated the aberration of light. This effect is due to the motion of the earth in its orbit and is an apparent change in the position of the fixed stars, relative to a frame of reference fixed in the earth, with a period of one year. Subsequently over a hundred years elapsed before any attempt was made to measure the velocity of light precisely.

Methods of measuring directly the velocity of light fall into two categories: (a) those which involve timing the passage of an electromagnetic wave over a known distance, and (b) those which involve measuring the wavelength λ of a wave of known frequency f, and deducing the velocity c from the simple relation $c = f\lambda$. Method (a) necessarily involves some sort of modulation of the wave, such as splitting it up into pulses, so that it carries a distinctive signal which can be used for timing; a wave of constant amplitude would clearly not serve. (Modulation other than amplitude modulation, although useful in principle in this context, has not been used and need not be considered.) The velocity which is measured is then the *group velocity* v_g, and is related to the phase (or wave) velocity v_p by

$$v_g = v_p - \lambda \frac{dv_p}{d\lambda}$$

The phase velocity is related to the velocity in free space c by

$$c = nv_p$$

where n is the refractive index of the medium in which the waves travel. The derivation of c from the measured group velocity therefore involves a knowledge of both the refractive index n and the dispersion $dn/d\lambda$ of the medium. The values of these quantities depend, in general, on the wavelength of the radiation used in the experiment so that the use, for example, of white light covering a wide range of wavelengths might lead to uncertainty in the correction. In dry air at atmospheric pressure and 0°C the correction from group velocity to free space velocity involves a factor whose value is close to 1·0003; the present uncertainty in the precise value of the correction is about ± 0·2 parts per million (provided that the atmospheric conditions are known). All the early work on the measurement of the velocity of light was performed in air, but the uncertainty in the correction to free space has always been small compared with the other uncertainties in the result. Of the three methods discussed in detail in this book only that of Bergstrand involves a correction of this sort.

Method (*b*) uses a continuous, unmodulated, wave and the phase velocity is measured. It has only been applied so far in the radio-frequency and microwave regions, since no techniques are available for the direct measurement of frequency in the infrared and visible regions of the spectrum. In the case of the method being used in a medium such as air, the only correction needed involves a knowledge of the refractive index and not its variation with wavelength. The method of Essen and Gordon-Smith is taken as an example of this type, but no correction was necessary as the measurement was made in a vacuum.

It is perhaps superfluous to point out that the measurement of a velocity must involve standards of both length and time, and a comparison of the experimentally determined quantities with the

EARLY WORK ON THE VELOCITY OF LIGHT

standards. Throughout much of the history of the measurement of the velocity of light the precision of the available standards and the precision of the available techniques for making comparisons with the standards has always been better than the precision of the experiment itself, though in the most recent determinations the length comparison makes a significant contribution to the final uncertainty. The time standard is invariably one of frequency, which is now the most precisely defined quantity available; frequencies may readily be compared with high precision.

The first terrestrial method used to measure the velocity of light was that of Fizeau (1849), who used a rotating toothed wheel to modulate a beam of light. The light pulses which passed through the openings in the wheel were reflected from a mirror at a distance of 8·6 km and returned through the wheel. At certain rotation velocities the returning light was cut off by a tooth, at others it passed through an opening, the wheel having rotated through an appropriate angle during the time of passage of the pulse of light to the mirror and back. The precision of the experiment depends upon the ability of the observer to judge the rotation velocity at which eclipse of the light, or maximum brightness, occurs, the former being the more easily judged. The precision is increased if a high rotation velocity is used so that a large number of teeth move past the point of observation during the outward and return journey of the light. Fizeau used a wheel with 720 teeth and rotation velocities of up to 200 rev/sec, a modulation frequency of the order of 100 kc/s. His result was 315,300 km/sec. The toothed wheel method was developed subsequently by Cornu (1876), Young and Forbes (1881) and Perrotin (1908). The last of these workers made a series of measurements between 1900 and 1902 and obtained a final result of 299,901 ± 84 km/sec.

An alternative technique for timing the passage of a pulse of light was suggested by Wheatstone (1834). A beam of light is reflected from a rotating mirror to a distant fixed mirror, back to the rotating mirror and then brought to a focus. If the mirror is stationary the final image appears at a certain place in the field of view of an eyepiece, but if the mirror rotates at a high enough angular velocity so that it has moved significantly during the time taken for the return passage of the light, the final image is displaced from its original position. In this form the technique is not capable of high precision since it requires the measurement of the displacement of the image, which in the earliest experiments of this sort was less than one millimetre. The technique was developed by Arago (1850) and Foucault (1850) and in 1862 Foucault obtained a value of the velocity of light of 298,000 km/sec (Foucault, 1862), though the uncertainty was about 500 km/sec. The baseline used by Foucault was only 20 m long, and clearly restricted the available precision, but it did enable him to compare the velocity of light in air with that in various liquids, and to show that they were in the inverse ratio of the refractive indices (though he did not realize at the time the difference between group and phase velocities in this connection). The method was eventually developed into one of high precision by Newcomb and Michelson.

Albert Abraham Michelson was born in Poland in 1852, emigrated with his parents to the United States, and there became one of the world's most distinguished experimental physicists. Most renowned for the experiment " On the relative motion of the earth and the luminiferous ether " (Michelson and Morley, 1887) he was the first American to be awarded, in 1907, the Nobel Prize in Physics. As an Ensign in the U.S. Navy, and an instructor at the Naval Academy at Annapolis, Maryland, Michelson became acquainted with the work of Simon Newcomb, a professor at the U.S. Naval Observatory, Washington, D.C., who was making a determination of the velocity of light by Foucault's method. Michelson was able to propose a modification of the optical arrangement which increased the brightness of the final image and allowed the use of longer baselines. His first measurements, made in 1878–9 over a baseline of 600 m, gave a result of $299,910 \pm 50$ km/sec (Michelson, 1879). A little later Newcomb (1880) suggested a further modification of the rotating mirror method which eliminated the measurement of the deflection of the image. The mirror was replaced by a rotating prism in the form of a precisely made regular polygon with polished faces. The rotation rate was adjusted until the returning light was reflected off the next following face of the prism: the image then appeared undisplaced. This " null " adjustment can be made precisely and, if the prism is sufficiently accurately made (which is readily achieved in practice), a knowledge of only the rotation velocity and the length of the optical path of the light is required. This method was used in principle by Michelson in all his subsequent determinations, and by Newcomb in 1881–2. Newcomb, working in Washington, obtained a result of 299,860 ± 30 km/sec (Newcomb, 1885), and in 1882 Michelson, then at the Case School of Applied Science, Cleveland, Ohio, obtained 299,853 ± 60 km/sec (Michelson, 1882).

Forty-one years later Michelson returned to the measurement of the velocity of light, with a lifetime of experience in experimental optics behind him. His object initially was to increase the length of the baselines used, relying on fundamentally the same timing technique as he had used in his 1882 measurements. An octagonal glass mirror was made with adjacent plane faces inclined to each other at angles which differed from 45° by less than one-tenth of a second of arc. The baseline was between Mount Wilson and Mount San Antonio near Pasadena, California, a distance of 35 km, which was measured by the U.S. Coast and Geodetic Survey to an accuracy of one part in a million. The rotation rate of the prism was compared stroboscopically with the frequency of a tuning fork, which in turn was measured by comparison with a standard invar pendulum. Later a steel octagon, glass and steel 12-sided mirrors and a glass 16-sided mirror were used. Michelson (1927) finally published a result of 299,796 ± 4 km/sec. He used a correction of + 67 km/sec to obtain the free space velocity from the measured velocity in the atmosphere, a value derived from the refractive index of the air, without taking into account dispersion. The extra correction involved, as Birge (1941) has pointed out, is + 2 km/sec, which is significant in view of Michelson's estimated error, and raises his result to 299,798 ± 4 km/sec. As it turns out, this result shows the best agreement of all those obtained by mechanical methods with the present-day value of 299,792·5 ± 0·3 km/sec, but, were the fact of this agreement not known, there would seem to be no special reason for selecting it as more trustworthy than the others; indeed, as is discussed later, there is a significant objection to it on the grounds of the instability of the baseline.

After his success with the 35 km baseline Michelson attempted measurements between Mount Wilson and Mount San Jacinto, a distance of 131 km, but these had to be abandoned because of the effect of atmospheric conditions, largely due to smoke from forest fires. He then turned his attention to the elimination of the atmosphere and the large correction which its presence introduces. A short baseline is inevitable if the experiment is to be done at low pressure, but a great accuracy is obtainable in the measurement of the baseline. The outcome was the mile-long evacuated tube measurement.

II
Michelson, Pease and Pearson

Interest in this paper is not confined solely to the impressive scale of the experiment. The final result had a profound effect on the accepted value for the velocity of light for over twenty years, yet it proved to be wrong by about 19 km/sec; its probable error was variously estimated at $\pm$ 4 km/sec and $\pm$ 11 km/sec. The source of this error, which was due to some systematic effect, has not been identified with any certainty.

Preparations for the measurement began in 1928 when Michelson, then at the University of Chicago, was invited by G. E. Hale of the Mount Wilson Observatory to return to California. The resources of the Observatory and the California Institute of Technology were placed at his disposal, and funds totalling $67,500 were provided by the Rockefeller Foundation, the Carnegie Corporation and the University of Chicago. Michelson took with him his technical assistant, Frederick Pearson, who had worked with him for several years, notably in connection with his Mount Wilson–Mount San Antonio measurements. His other collaborator, Francis G. Pease, was an astronomer and designer of optical instruments at Mount Wilson Observatory who had previously worked with Michelson: he was the first to use Michelson's stellar interferometer to measure the angular diameter of a star, and had taken part with Michelson in a repetition of the Michelson–Morley experiment. Their paper opens with a brief explanation of the decision to undertake the measurement in a form distinct in two respects from the Mount Wilson–Mount San Antonio determination. Emphasis is placed on the higher accuracy of the measurement of the shorter baseline, and the desirability of removing the air from the path of the light "... giving a small well defined image, unaffected by atmospheric disturbances ...". Neither of these hopes was realized. Although the results of the measurements of the baseline by the U.S. Coast and Geodetic Survey are quoted to one-tenth of a millimetre in 1·6 km, the results show a total variation of 13·1 mm over the period March 1931 to February 1933, and then a change in the opposite direction of 8·5 mm between February 1933 and July 1933; an earthquake had occurred in March 1933. Although 13·1mm corresponds to a variation of 1 part in 122,000, it is far too small to account for the final error of nearly 1 part in 15,000. Nevertheless, it shows the unfortunate choice of location in a region of the world susceptible to earthquakes and, no doubt, less obvious earth movements about which little information exists. It casts doubt on Michelson's previous attempt over the 35 km baseline, since, as Miller has pointed out, a major earthquake occurred at Santa Barbara, 150 km from Mount Wilson, between the measurement of the baseline in 1922 and the subsequent determinations of the velocity of light in 1926. At the time Michelson estimated that the baseline length was known to 1 part in a million, and placed an uncertainty in his final result of 1 part in 75,000. This, for measurements between mountain peaks, is probably optimistic in the light of the observations on the changes of baseline length over a short distance of level terrain in the neighbourhood (Santa Ana is about 60 km south of the Mount Wilson–Mount San Antonio baseline).

The use of an evacuated pipe certainly reduced the effect of the refractive index of the atmosphere to a negligible factor (1·2 parts in a million) but introduced difficulties in connection with the achievement of a clear image of the slit. The light traversed the pipe eight times for the majority of the measurements, and the method was basically the Newcomb modification of the Foucault method, described above, to adjust the mirror rotation rate until the image of the slit appeared undisplaced, using a 32-sided prism with mirror faces. In practice the rotation speed of the mirror was held constant at a value, in revolutions per second, exactly twice that of the vibration of a tuning fork. (For the early experiments of the series the mirror speed and the fork frequency were equal.) The rotation speed was, in these circumstances, not quite that necessary to bring the image of the slit (Fig. 1.1 of the paper) to the same position it occupied when the mirror was stationary, and the resulting small displacement was measured. The need for a zero reading was eliminated by taking readings for opposite directions of rotation of the mirror. The total distance d between image positions was only a few thousandths of an inch; an error of 0·001 in. in this measurement would have resulted in an error of 30 km/sec in the final results, so that it is clear that a sharp undistorted final image was of paramount importance.

The difficulties encountered in obtaining a good image make sad reading. It is evident that the use of a tube set up far larger temperature gradients in the remaining air than were encountered in the free atmosphere, and these gradients distorted the beams to such an extent that good working conditions were the exception. The pressures attainable were between 1·5 and 0·5 mmHg, of the order of 0·001 atm; the residual gas, if one assumes it to have been air, had a refractive index of about 1·0000003 which would have changed by only 10^{-9} for each degree centigrade change of temperature. The form and magnitude of the temperature gradients which existed in the tube are, of course, a matter of conjecture, but, in spite of the very small changes of refractive index involved, could readily have accounted for a net displacement of the beam over the light path sufficient to cause the disappearance of the image. The best images, obtained " when a quiet fog settled round the tube ", seem to be related to a uniform tube temperature, no temperature gradients and, consequently, no turbulence of the remaining air in the tube. The comparatively high-pressure limit of the available vacuum pump was unfortunate, but at that time techniques were not readily available for evacuating such a large tube to a pressure several orders of magnitude lower. Modern diffusion pumps used on high energy accelerators, such as proton synchrotrons, produce pressures below 10^{-4} mmHg which would have been ample to eliminate the effect of the residual gas, though the sealing of the joints in the pipe might have had to be improved to reduce leakage. The whole installation would, if improvements of this sort could have been undertaken, have been very much more costly.

During the building of the apparatus and the early observations Michelson was ill; on 9 May 1931 he died. The majority of the observations which contributed to the final results were made by Pease and Pearson after his death. The histogram in figure 1.8 of the paper summarizes the measurements, which appear at first sight to lie close to a Gaussian distribution such as would be expected if the errors of measurement had a purely random character. The number of separate sets of observations from which a mean value for the velocity of light was derived was 233; each of these was given a " weight " equal to the number of separate determinations within the set. The total number of individual determinations was 2885·5 (the 0·5 accounts for an incomplete determination). Applying statistical analysis to the distribution gives a weighted mean velocity of 299,773·85 km/sec with a probable error of 0·17 km/sec. The latter figure is, however, quite meaningless in this case, and the application of such an analysis is totally unjustifiable. As Worthing and Geffner (1944) have pointed out, the distribution departs considerably from a normal distribution; but there is more to the objection than this. Examination of the sequence of Pease and Pearson's observations set out in Table VI shows that the distribution of the individual results about the mean of 299,774 (in round figures) is far from random. The average deviation (the arithmetic average of the deviations of the individual observations from the mean) of all the observations is 11 km/sec; if all the observations had a random error one would expect them to be randomly distributed about the mean when considered in the order in which they were taken. This is far from the case, as the authors themselves point out. Among the earlier readings for example, there is a group taken between 25 March and 3 April 1931 which has an average value widely different from the final result: the average value is 299,746 km/sec and their average deviation 12 km/sec. Throughout the later sets of observations there are similar cases of variations of the average value, sometimes with a pronounced periodicity. Some systematic effect was causing these fluctuations, and was in all probability itself responsible for the error of the final result. In other words, a systematic error was always present, but was itself varying in magnitude. The authors make an attempt to correlate the variation of the results with the local tides, and with the diameter of the moon, but the relationship is by no means obvious.

Instability of the terrain has been suggested as a cause of the fluctuations and of the final error of 19 km/sec. This corresponds to an error of nearly 10 cm in the 1·6 km baseline; the four measurements of the baseline showed an overall variation of 1·31 cm in over two years. It is possible, but extremely unlikely, that there were variations between the measurements of the baseline which were not present when those measurements were made, but a total variation of almost 50 cm in the baseline is necessary to account for the difference between the results of 299,728 ± 13 km/sec on 31 March 1931 and 299,821 ± 12 km/sec on 9 March 1932. Such a large change in the baseline is hardly credible. The source of error must be sought elsewhere.

The linear displacement of the slit image in the focal plane of the eyepiece (Fig. 1.1) when the mirror was rotating in synchronism with the tuning fork was typically (Table IV) 0·0057 in.; this is the total displacement when the mirror rotation direction was reversed. An error of 0·001 in. in this quantity corresponds to an error in the final result of nearly 30 km/sec. The accuracy of the measurement is evidently particularly sensitive to the position of the slit image. Because of the method of measuring the displacement with the mirror rotating in opposite directions one has to find an explanation for a spurious displacement which reverses in direction with the reversal of the mirror rotation. It is relatively easy to do so: Bergstrand (1952) has pointed out that the atmospheric conditions in the vicinity of the rotating mirror must have been considerably disturbed by the presence of the air from the jets driving the turbine situated immediately below the rotating mirror (Fig. 1.3). (Michelson's earlier paper (1927) gives an instructive photograph of the type of arrangement used.) A local change of the refractive index of the air by a combination of temperature and pressure gradients would have deflected the light if it travelled at an angle to the surfaces of equal refractive index. Such a deflection would have been irrelevant if it had occurred equally in the beams incident on the rotating mirror before and after the light travelled over the long optical path in the tube; but the arrangement was asymmetrical, the outgoing light was reflected from the lower part of the mirror faces and the returning light from the upper part. The turbine was below the mirror, so that the effect of the escaping air was almost certainly not the same for the two beams. The direction of air flow, and consequently the sense of any deflection of the light beams which might have occurred, reversed when the direction of rotation of the mirror was changed. Only one mirror and one turbine were used throughout the measurements. The variation of the systematic error might be explained by the different rates of air flow needed to maintain the turbine speed due to slight variation from time to time in the frictional resistance of the turbine bearings, which were lubricated with oil. In the Mount Wilson–Mount San Antonio determination six different mirrors were used, and two different turbines. There is far less evidence for a temporal variation of the results (which were, however, confined to the period June–September 1926) and the symmetrical orientation of the outgoing and incoming beams in this experiment make it far less likely that there was an error introduced by the presence of pressure or temperature gradients in the air near the mirror.

A similar source of error would have been an angular displacement of the 90° prism I (Fig. 1.1) which changed in magnitude in direction with the direction of rotation of the turbine. This point seems to have occurred to the experimenters (p. 49) though no details are given of the sensitivity of the check that was carried out. Alternatively, the rotation rate of the mirror might not have been uniform during one revolution; it is assumed that the time taken to turn 1/32 of one revolution was precisely 1/32 of the measured period. The mirror was, however, driven by the turbine, which was subjected to periodic impulses as the depressions in the turbine wheel passed by the air jets. In the absence of any data on how long the turbine took to slow down when the air jets were turned off it is not possible to make an estimate of the effect.

The detection and elimination of systematic errors is the most difficult, yet the most important, feature of the field of precise measurements. The periodic fluctuations in the measured value of the velocity of light were clearly of concern to the investigators yet they give no indication that they made any attempt to eliminate their source. In the absence of tests which might have revealed the cause of the fluctuating error, a complete re-design of part of the apparatus, such as the turbine, might have eliminated the systematic error and yielded results with a random time distribution of deviations from the mean, in which greater confidence could have been placed. Particularly disturbing is the statement " The fact that these mean results (299,775 km/sec and 299,746 km/sec) differed from each other and from the value 299,796 km/sec obtained from Mount Wilson necessitated additional readings ". It sounds as though the authors were about to adopt the practice of taking measurements until agreement was reached with previous work. As it turned out, this procedure would have proved fruitless, as the mean result was consistently lower than the Mount Wilson value, and they contented themselves with taking a large number of observations with the apparatus originally set up.

The paper gives an exceptionally detailed account of most of the experimental details, the procedure during observations, and the results.

III
The Period 1935-46

The Michelson, Pease and Pearson experiment was the last to use a mechanical device for modulating or deflecting a beam of light. The desirability of the use of a higher frequency in order to reduce the length of the light path, or to achieve greater precision over a long path, led to the use of electro-optical devices for modulating a light beam at much higher frequencies than could be achieved mechanically. Two types of modulators were used, the electro-acoustic diffraction grating and the Kerr cell. The former consists of a piezo-electric crystal, such as quartz, in which standing acoustic waves are generated by applying a high frequency alternating electric field. Light is passed perpendicularly to the direction of propagation of the acoustic waves, and, because of the variations of refractive index arising from the compressions and refractions in the sound wave, the optical path through the crystal varies accordingly. The crystal thus acts as a diffraction grating in which the phase, rather than the amplitude, of the wavefront varies in a regular manner. This diffraction grating is formed and disappears at twice the acoustic frequency. The diffracted light is thus amplitude modulated. Although this type of modulator has been used by two workers, McKinley (1950) (the experiments were done in 1937-8) and Houston (1938, 1941, 1950), their results for the velocity of light are not of significant value. McKinley used a frequency of 8 Mc/s, and Houston 200 Mc/s.

The Kerr cell uses the fact that some liquids, notably nitrobenzene, become optically active when placed in an electric field, the optic axis coinciding with the field direction. Plane polarized light passed perpendicular to the field becomes, in general, elliptically polarized. Such a device placed between crossed polarizers is capable of modulating the intensity of a beam of light and can be used at frequencies up to between 10 and 100 Mc/s. The Kerr cell has been used by Gutton (1912) to compare the velocity of propagation of light with that of waves on a transmission line, and by Karolus and Mittelstaedt (1928, 1929) to measure the velocity of light. Subsequent determinations of the latter quantity were made by Hüttel (1940) and Anderson (1937, 1941). Hüttel used modulation frequencies between 5 and 12 Mc/s and a light path of 80 m, Anderson a path of 172 m and a frequency of 19 Mc/s.

The measured values of the physical constants had been critically reviewed by Birge from time to time since 1929, and when the results of these determinations of the velocity of light became available he undertook (Birge, 1941) a detailed analysis leading to the most acceptable value for the velocity. To the result of Michelson, Pease and Pearson's determination he assigned a probable error of $\pm$ 4 km/sec, although their own estimate was $\pm$ 11 km/sec. Anderson published a final result of 299,776 km/sec (compared with 299,774 km/sec obtained by Michelson, Pease and Pearson) and Birge estimated his probable error to be $\pm$ 6 km/sec. Hüttel's result was 299,768 $\pm$ 10 km/sec, which Birge corrected (using the group, rather than phase, refractive index for air) to 299,771 km/sec. The greatest weight he gave to Michelson, Pease and Pearson; earlier measurements, except that of Rosa and Dorsey (discussed later), were scarcely considered. His final result for the most acceptable value for the velocity of light was 299,776 $\pm$ 4 km/sec.

This value, which we now believe to have been in error by some 16 km/sec, was in agreement with his 1934 recommendation, and was consequently accepted and used for nearly twenty years. The acceptance of Birge's value was based on the agreement between the results obtained by Michelson, Pease and Pearson, by Hüttel and by Anderson, using two different techniques and working completely independently. The systematic errors present in these three determinations may, by chance, have led to agreement between the final results. Alternatively Bearden and Thomsen (1959) have suggested ". . . there was a subconscious psychological factor which tended to make each experimenter look for errors in his technique until he could check the then accepted value ". Such an effect, unfortunate as it is, may well have contributed to the agreement between the results of an experiment conducted on an impressive scale on the initiative of an accepted master such as Michelson, and those of subsequent workers using new and unfamiliar techniques.

World War II brought determinations of the velocity of light temporarily to a halt, but produced important new advances in methods of microwave generation. By the end of the war the development of radar had resulted in the availability of sources of radiation, both continuously operating with modest power output (the klystron), and high-powered and pulsed (the magnetron), at wavelengths down to the region of 1 cm. Radar itself provided the first evidence of the erroneous value of the accepted value of the velocity of light, though it appears that this evidence was, at the time, ignored. An airborne radar navigation aid, Shoran, is based on the principle of timing the passage of a short pulse of radio waves from an aircraft, to a ground station, and back; it was used by Aslakson (1951) to measure the distance between two ground stations about 160 km apart, the distance having already been measured geodetically. Aslakson had by 1947 obtained sufficient evidence (Aslakson, 1949) to question the Birge value of the velocity of light, and his work supported the results of the microwave cavity measurements which were being made at that time by Essen and Gordon-Smith at the National Physical Laboratory. Only then were the doubts about the Birge value taken seriously. Very soon afterwards confirmation came from optical measurements made by Bergstrand over a long path using a Kerr cell modulator.

IV
Essen and Gordon-Smith

The principle of the method adopted by these workers is basically the measurement of the wavelength of microwave radiation whose frequency is known. Since the velocity of electromagnetic radiation *in free space* is required, the wavelength also must be that in free space. The direct determination of the free space value of a microwave wavelength to an accuracy of the order of one part in a million is difficult because the wavelength is modified not only by the refractive index of the medium in which the waves are propagated, but also by the proximity of conductors or dielectric boundaries. This latter difficulty does not arise in optical determinations since the wavelength is small and any of the relevant dimensions of the apparatus (for example, the diameter of the vacuum tube in Michelson, Pease and Pearson's measurement) is clearly a very large multiple of the wavelength and so has a completely negligible effect. The extreme example of the influence of boundaries on wavelength is provided by the propagation of microwaves in a waveguide, in which case the lateral dimensions of the waveguide are comparable with the wavelength. Here the free space wavelength λ is related to the wavelength in the waveguide λ_g to a good approximation by the relation

$$\frac{1}{\lambda^2} = \frac{1}{\lambda_g^2} + \frac{1}{\lambda_c^2}$$

where λ_c, the critical wavelength, is a parameter which depends on the lateral dimensions of the waveguide. The wavelength in the guide is thus always greater than the free space wavelength, corresponding to a phase velocity greater than the free-space velocity. When the free-space wavelength is greater than the critical wavelength of the guide the wave is rapidly attenuated and free propagation along the guide is not possible. The relation between the phase velocity v_p and the group velocity v_g is $v_p v_g = c^2$, and the guide is dispersive.

The above considerations have to be modified if the guide is filled with a medium (such as a dielectric) but such a case is unusual and need not be considered here.

A length L of waveguide is a resonant cavity for radiation of free-space wavelength λ if $L = n\lambda_g/2$ where n is an integer. The critical wavelength λ_c is known from theoretical considerations, so that λ can be found from the measured value of L and the lateral dimensions of the cavity. If the frequency f of the radiation at resonance is known, the free-space velocity v_0 of the radiation can be found simply from $v_0 = f\lambda$. The accuracy of the measurement clearly depends on the measurement of the cavity dimensions and the determination of the resonant frequency. The latter in turn depends upon the quality factor Q of the cavity, since the width of the resonance curve of the cavity under conditions of constant excitation is inversely proportional to Q (a precise relation is given on p. 91).

In designing an experiment of this sort it is possible to choose a fixed frequency and alter the length of the cavity until resonance is obtained, or to use a cavity of fixed dimensions and alter the frequency to find the resonance. The latter method was chosen by Essen and Gordon-Smith; the construction of the cavity is simpler and the measurement of its dimensions more straightforward; Essen (1950) later adopted the former method and, in effect, measured the half-wavelength in the cavity by finding the cavity length for successive resonances at a fixed frequency, thus eliminating some end-effects. In either case a precise measurement of the oscillator frequency was needed. The frequency was in the region of 3000 Mc/s ($\lambda = 10$ cm approx.) and was measured in terms of the frequency of a quartz crystal oscillator operating at 100 kc/s. The frequency of this oscillator was known to be better than 1 part in 10^8, and by successive stages of frequency multiplication a number of harmonically related frequencies were generated in the region of 3000 Mc/s. The output of the oscillator at an unknown frequency was heterodyned with one of the precisely known frequencies and the resulting difference frequency was determined by comparison with an interpolation oscillator of known frequency. In this way the microwave oscillator frequency could be determined with a precision of $\pm$ 2 parts in 10^6. The cavity used by Essen and Gordon-Smith was in the form of a right circular cylinder, and the modes of excitation were E_{010}, E_{011},

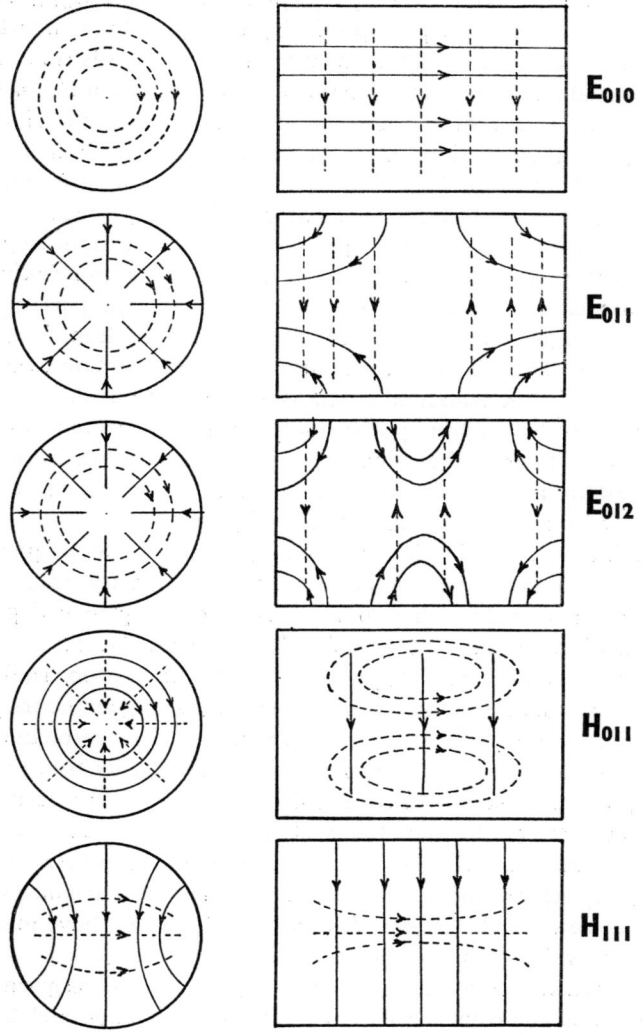

FIG. 4.1. The resonant modes of a cylindrical cavity used by Essen and Gordon-Smith.

——— Electric field; - - - - - - Magnetic field.

E_{012}, H_{011}, and H_{111} (an equivalent notation is TM_{010}, TM_{011}, TM_{012}, TE_{011} and TE_{111} respectively), though only the first two of these were used in the final measurements. The distributions of electric and magnetic fields in these modes are shown in Fig. 4.1. The E_{010} mode is a special case, which does not correspond to any waveguide transmission mode, in which the electric field intensity is uniform in the axial direction, and the resonant frequency is independent of the length of the cavity. The cavity dimensions were chosen so that the resonant frequencies of the above modes lay in the region between approximately 3000 and 5000 Mc/s in which klystrons were available. The cavity was excited by inserting a probe in an appropriate fashion into one end of the cavity: the E modes require a straight axial probe, and the H modes either a loop, or a wire bent so as to excite an azimuthal component of electric field. Resonance was detected by a second probe similar to the excitation probe. The presence of these probes obviously distorts the field distribution to some extent, and it was important to ensure that their effect on the calculated resonant frequency was negligible.

An important point in determining precisely the frequency at which the peak of the resonance curve of the cavity occurs is the elimination of errors due to resonances or frequency dependent transmission factors in the associated equipment. Thus if the amplitude of the voltage which appears on the excitation probe is, because of some resonance in the connecting cables, falling as the frequency of excitation increases, the apparent position of the resonance peak would be displaced to the low frequency side of its true position. This point is discussed in section 4 of the paper. The cavity itself was evacuated to eliminate the need for correction due to the dielectric constant of the air; no question of the distinction between phase and group velocity arises, but, had the cavity contained a medium the measurement would, as has already been discussed, have measured the phase velocity in the medium.

As Essen and Gordon-Smith point out, the expressions given above assume that the cavity walls are perfectly conducting, and that the unloaded Q of the cavity is infinite. (The " unloaded Q " is the quality factor of a cavity which is not coupled to any external load capable of absorbing energy from the cavity.) For a cavity of finite Q the resonant frequency is reduced by a factor $[1 + (1/2Q)]^{-1}$.

The values of Q obtained in practice lay between 10^4 and 2×10^4, so it is clear that the correction was of importance in an experiment aimed at a final error of the order of 1 in 10^5. The Q of the cavity was measured by finding the frequency width of the resonance curve of the cavity. The observed Q was, in fact, the loaded Q of the cavity, since the input probes extracted power from the cavity, but as the authors indicate, the fraction of the power loss due to this cause was negligible with the probe settings used, so that the loaded Q and unloaded Q were identical; the only significant power loss was due to conduction currents flowing in the walls, and the theoretical correction could be applied. The justification for the use of the correction term was, when the paper was written, solely on theoretical grounds, as set out in section 2 of the paper. Later work by Essen (1950) using a cavity of variable length in several different modes, and at two different frequencies, showed that the uncertainty in the magnitude of the correction term gave a maximum error in the final result of a little over 1 km/sec. Essen's later result, based on this more thorough investigation of the applicability of the correction term, was 299,792·5 $\pm$ 3 km/sec, against Essen and Gordon-Smith's result of 299,792 $\pm$ 9 km/sec.

In both cases the quoted uncertainty is the maximum possible error, obtained by adding the numerical values of the estimated errors from all sources. For this reason the accuracy of the measurements appears lower than can be justified when they are compared with other workers' results. The paper of Essen and Gordon-Smith was the first to describe in detail work which revealed the error in Birge's value for the velocity of light, and subsequent measurements have fully supported the value that Essen and Gordon-Smith obtained. Essen's estimate of the probable error of the earlier result is $\pm$ 3 km/sec, and of his later result $\pm$ 1 km/sec.

Comparison of the quoted error limits of results obtained by different workers is always difficult, even when their method of deriving the limits is given in detail. In all cases there is the possibility of the existence of an undetected systematic error. Often statistical methods are applied to a small number of results, so that the magnitude of the derived probable error has itself a considerable uncertainty. Very few workers quote their estimated maximum possible errors and in other cases the meaning of the quoted errors is often obscure. The "probable error" should mean that there are equal probabilities of the true value lying between the quoted limits as outside them but in most cases where such limits are given their value depends upon the experimenter's judgement rather than the strict application of statistical analysis. Consequently the maximum possible error limits, outside which it is extremely unlikely that the true value lies, have a more definite interpretation than other forms of quoted error. Essen and Gordon-Smith's determination, supported by Essen's later work using a similar method, can be regarded as the first of a group which were made with high precision and which are essentially in agreement: the others are Bergstrand, using visible light over a comparatively long path, and Froome, using a microwave interferometer in air.

V
Bergstrand

The inception of Bergstrand's measurement of the velocity of light is explained at the beginning of his paper: the object was to develop an apparatus for the measurement of distances as an alternative to conventional geodetic surveying methods. The outcome of his work was the " Geodimeter ", a portable instrument manufactured by the Aga Company (Svenska Aktiebolaget Gasaccumulator) of Stockholm. Distances are measured with this instrument by the same technique as Bergstrand used for the velocity of light determination, but the velocity of light is treated as a known quantity. Bergstrand's measurement was intended to establish a value for the velocity of light suitable for the subsequent users of the instrument. His value is entirely supported by independent measurements, and is quoted to an accuracy of $\pm$ 0·25 km/sec, or better than 1 part in a million. In a paper published after the one reprinted here Bergstrand (1951) reported a repetition of his earlier work using a different baseline and gave an unchanged value for the velocity, 299,793·1 km/sec and a slightly reduced error of $\pm$ 0·2 km/sec.

The principle of the method is basically that used by Hüttel (1940). A light beam is sinusoidally amplitude-modulated by passing it through a Kerr cell. The light then travels over a known path and back to a detector, where the phase of the modulation is measured relative to the phase of the modulation of the outgoing light. In this simple form the measurement would not be capable of high precision for two reasons: the uncertainty in the length of the light path (bearing in mind that it must be known to be considerably better than $\pm$ 1 cm), and the difficulty of making a precise determination of the relative phase of two sinusoidal signals (the phase would have to be determined to the order of $\pm$ 1/10 degree at a modulation frequency of about 10 Mc/s). Bergstrand's apparatus eliminates the first uncertainty by using two light paths, one short and one long; the difference between the two paths, which can be precisely determined, is the important quantity. The second source of error is reduced by making the instrument a very sensitive detector of phase by a null point method, so that the need for a measurement of some arbitrary relative phase is eliminated. The lengths of the two light paths are adjusted so that the instrument gives a null reading for both. The difference in the light paths is then known to be a multiple of the modulation wavelength of the light, and the velocity is deduced from the known difference in the light path and the modulation frequency, with an appropriate correction for the effect of the air in the light path.

Bergstrand describes the action of the Kerr cell modulator and the null-detector in detail. The principle of the method is as follows. The Kerr cell is supplied simultaneously with a potential ("tension") of amplitude 2000 V at a frequency in the region of 8·33 Mc/s, and a 50 c/s square wave with an amplitude of 5000 V. The light transmitted by the Kerr cell varies with potential difference in the manner shown in Fig. 5.1(a). When the 50 c/s voltage is positive, the light transmitted is as shown in Fig. 5.1(b), when it is negative it is as shown in Fig. 5.1(c). There is a phase change of 180° of the transmitted light relative to the phase of the 8·33 Mc/s voltage applied to the Kerr cell. The returning light, after traversing the chosen light path, falls on a photomultiplier connected to an amplifier. The supply voltage to the final collector or anode of the photomultiplier, which in normal use has a steady positive value, is in this case the same alternating voltage used to supply the Kerr cell. The photomultiplier is completely insensitive when the anode supply voltage is negative, and has almost constant sensitivity during most of the positive half-cycle, as indicated in Fig. 5.2(a). When the light falling on the photomultiplier has the phase shown in Fig. 5.2(b) (corresponding to the part of the 50 cycle Kerr cell voltage shown in Fig. 5.1(b)) the mean photomultiplier current is large, because high light intensity corresponds to the sensitive half-cycles of the photomultiplier, but when the phase is changed by 180° (Fig. 5.2(c), corresponding to Fig. 5.1(c)) the low intensity half-cycles correspond to the sensitive photomultiplier half-cycles and the mean photomultiplier current is small. When the length of the light path has a value which gives some intermediate relative phase of the light intensity and the photomultiplier sensitivity, the mean photomultiplier current has

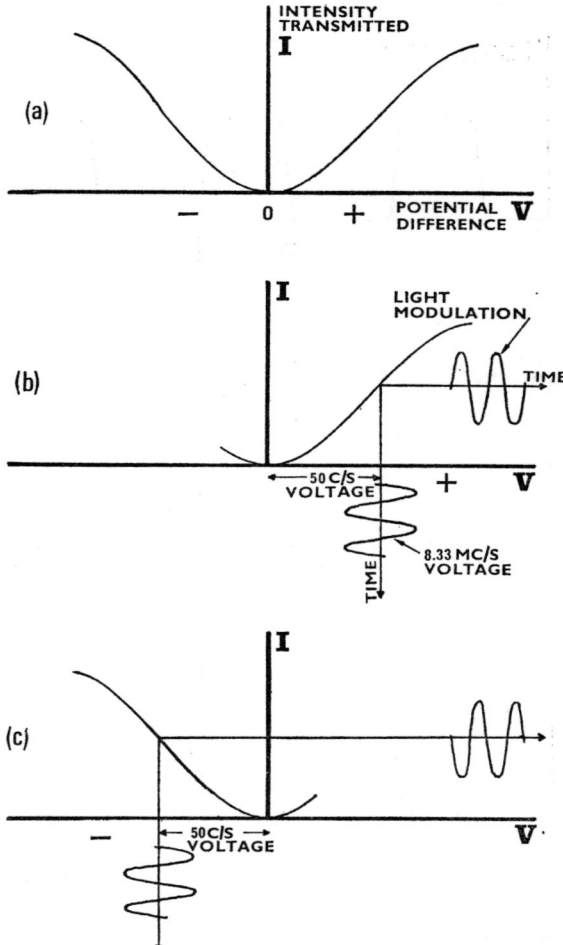

FIG. 5.1. (a) Light transmitted by a Kerr cell between crossed polarizers when a potential difference V is applied to the cell. (b) Amplitude modulated light produced by applying simultaneously a 50 c/s and an 8·33 Mc/s alternating potential to the Kerr cell; 50 c/s potential positive. (c) as (b) but with 50 c/s potential negative; note the 180° phase change of the 8·33 Mc/s modulation compared with (b).

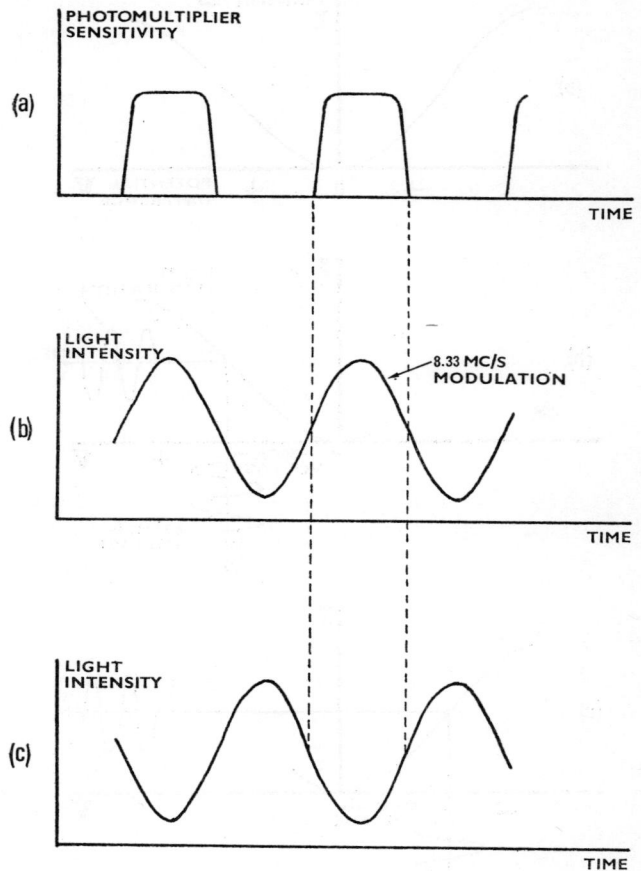

FIG. 5.2. (a) Variation of photomultiplier sensitivity with time; repetition rate 8·33 Mc/s. (b) Amplitude modulated light incident on photomultiplier with a phase which gives maximum mean photomultiplier current. (c) As (b), but with 180° phase difference, giving minimum mean photomultiplier current.

some intermediate value, as shown in Fig. 5.3. Since the phase is varied by 180° every 1/100 second, the mean photomultiplier current varies, in general, from a value such as *A* to a value *B*, where these points are separated by a phase difference of 180° in Fig. 5.3. The photomultiplier current therefore has, in general, a component at 50 c/s. If the phase of the returning light is such that the two points separated by 180° are *C* and *D*, there is no 50 c/s component in the

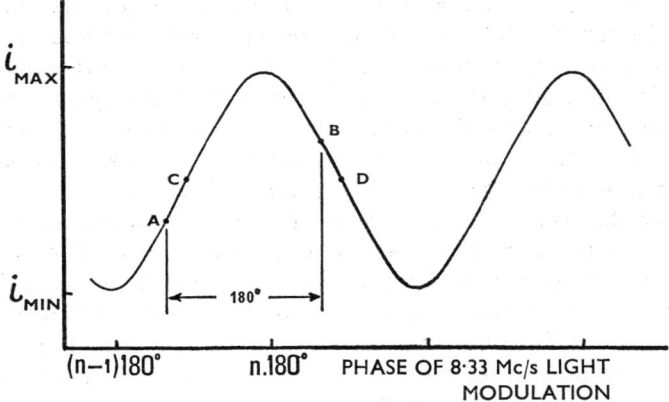

FIG. 5.3. Variation of mean photomultiplier current with the phase of the modulation of the incident light; i_{max} and i_{min} correspond to Figs. 5.2(*b*) and 5.2(*c*).

current. The points *C* and *D* can only correspond to phase differences of $(n \pm \frac{1}{2})180°$, where *n* is an integer, and can be found very precisely by using a sensitive circuit which gives a null reading when there is no 50 c/s component in the photomultiplier current. This circuit is formed by the valves 1 and 2 of Fig. 3.2, with the indicating meter I.

Successive null points, as the light path is varied, are separated by a phase difference of the modulation envelope of the light of 180°. The modulation frequency of 8·33 Mc/s corresponds to a modulation envelope wavelength of close to 36 m, so that successive null points appear for positions of the distant mirror separated by 9 m approximately. Bergstrand has stated that the sensitivity of the method was such that a movement of the distant mirror of 4 mm could have been detected.

The baseline used in the experiments described by Bergstrand was close to 7 km. The instrument itself was, obviously, fixed in position, and it was inconvenient to move the distant mirror until a null, or zero-point, reading was obtained on the meter. To overcome this, the modulation frequency was altered slightly until a zero reading was achieved. The shorter light path was built into the instrument itself, and was variable in length, as shown in Fig. 3.1. Bergstrand describes how the difference in these two light paths was measured to an accuracy of a few mm in nearly 7 km. By using this technique the effect of any time delays in, for example, the response of the photomultiplier were eliminated (provided that they were constant during a determination). The systematic error in the result obtained by Anderson (1941) may have arisen from this cause.

Near the end of the section on the general description of the apparatus Bergstrand points out a systematic error due to the different intensity of the light passed by different parts of the Kerr cell, which would have given a spurious 50 c/s signal from one light path and not from the other (because different parts of the light beam were used in the two cases) when in both cases a zero point should have been indicated. The error was eliminated by reversing the phases of both the high frequency supply to the Kerr cell (SwI of Fig. 3.2) and the 50 c/s supply to the detector circuit (SwII) and taking the mean of the four null settings (of the light path or the oscillator frequency, depending upon whether the short or long light paths respectively were being measured) for the four possible combinations of the switch settings.

The success of Bergstrand's determination lies in the sensitivity of the method and the elimination of sources of systematic error. The uncertainty in the final result arises from a number of causes, all of about the same magnitude: the baseline measurement provides the largest; the zero-point settings for the near and distant mirrors, the correction for the presence of the air (partly due to uncertainties in atmospheric conditions and partly to the uncertainty in the effective wavelength of the light) and the frequency of the oscillator also contribute. Bergstrand combines these uncertainties by taking their root-mean-square value. The individual velocity measurements over the 7 km baseline, thirteen in number and taken in two groups in the spring and autumn of 1949, show an overall spread of only ± 0·885 km/sec. Detailed results for all these measurements are given in the tables. The standard deviation of the mean is found by a standard statistical procedure based on the assumption that all the deviations from the mean are random, with an appropriate weight for each determination which depended on the number of individual observations which had contributed to that determination. Some additional measurements were made over shorter sections of the baseline, and earlier measurements were re-examined and corrected. The final result for the velocity of light is the result of combining all these measurements, but those over the 7 km baseline are by far the most significant. The final quoted value is 299,793·1 ± 0·25 km/sec.

VI
More Recent Measurements

The Geodimeter has been used since 1950 on several occasions over different baselines to measure the velocity of light. Bergstrand himself (1951) obtained a value of 299,793·1 ± 0·2 km/sec by combining his earlier measurement with a new determination over a 5·4 km baseline on the island of Öland. Mackenzie (1954) in Great Britain obtained a result of 299,792·3 ± 0·5 km/sec, Schöldström (1955) in Sweden obtained 299,792·4 ± 0·4 km/sec and Waller (1956) in Australia, 299,792·5 km/sec. The use of the instrument over independently measured baselines, with a variety of atmospheric conditions, and using independently calibrated oscillators, reduces considerably the possibility of the presence of a systematic error. Unless there is a common source of systematic error which has hitherto been overlooked the results of the Geodimeter measurements may be accepted with confidence. Bergstrand (1957) has reviewed all the Geodimeter measurements and from them deduces a value for the velocity of light of 299,792·85 ± 0·16 km/sec. In 1957 a value of 299,792·5 ± 0·4 km/sec was recommended by the 12th General Assembly of the International Scientific Radio Union, and accepted by the International Union of Geodesy and Geophysics.

The use of microwaves was continued at the National Physical Laboratory after Essen's work by Froome, who developed a microwave interferometer capable of high precision. In the various forms of this instrument (Froome, 1952, 1954, 1958) microwaves at wavelengths of 1·25 cm or 4 mm were radiated from a horn, and their wavelength was measured by an interference method. The apparatus was situated (in air) in a large room. The wavefronts were not plane, and, in the determination of the wavelength, corrections were necessary to take into account the shape of the wavefronts and the refractive index of the air. Froome's final result was 299,792·5

± 0·1 km/sec; the quoted uncertainty is a standard deviation. This result is therefore in good agreement with the Geodimeter measurements and Essen's microwave cavity determination. Not only does the agreement lead to confidence in a value for the velocity of light close to 299,792·75 km/sec, with an uncertainty of no more than about one part in a million (or ± 0·3 km/sec), but it shows that within a limit of this order of magnitude there is no difference in the velocity at a frequency of the order of 10^{11} c/s used by Froome, and the velocity at a frequency in the visible region, of the order of 10^{14} c/s.

Measurements of the velocity of electromagnetic radiation have been made outside the frequency limits just quoted, but with far less precision. Cleland and Jastram (1951), and Luckey and Weil (1952) found the velocity of gamma radiation (with frequencies in the range 10^{20}–10^{22} c/s) to be 298,300 ± 1500 km/sec and 297,400 ± 3000 km/sec respectively. In the radio frequency region Florman (1955) found the velocity of 172·8 Mc/s waves to be 299,795·1 ± 3·1 km/sec. Maxwell's theory predicts that the velocity of electromagnetic radiation is independent of frequency, and there is at the moment no experimental evidence to the contrary.

INDIRECT METHODS

It was pointed out in the Introduction that, according to Maxwell's theory, the velocity of light is identical with a constant that appears in the ratio of the electrostatic to the electromagnetic system of units. A value for the velocity of light can thus, if one accepts the theoretical premises, be found in an experiment which compares the two systems of units. The measurement which is particularly suited to this purpose is the evaluation of the capacitance of a condenser in the two systems. In the electrostatic system the capacitance can be calculated from the physical dimensions of the condenser; clearly only simple geometrical shapes lend themselves to an accurate calculation. The capacitance in electromagnetic units can be found in principle by placing a charge, known in e.m.u., on the capacitor and measuring the resulting potential difference in e.m.u. A more practicable method is to arrange the condenser in a circuit which alternately charges and discharges it so that it passes a mean current which is calculable in terms of the charging potential, the frequency of the charge–discharge cycle and the capacitance of the condenser.

A very elegant experiment of this sort was done by Rosa and Dorsey at the National Bureau of Standards, and is described in detail by them in a 171-page paper (Rosa and Dorsey, 1907). They used condensers in the form of concentric spheres, coaxial cylinders and plane parallel plates, the last two with guard rings. The capacitances were found from the dimensions, and by placing a condenser–commutator combination, as described above, in one arm of a Wheatstone bridge. This combination was equivalent to a resistance which was compared with the value of the International Ohm (defined in terms of the resistance of a specified column of mercury) to the Absolute Ohm (defined in terms of the ampere, the joule and the second, and hence in electromagnetic units). The ratio of these two values for the ohm has to be found separately. The most recent evaluation gives, using Rosa and Dorsey's data, a value of the velocity of light of 299,800 km/sec (Silsbee, 1955). Birge (1941) assesses the uncertainty of Rosa and Dorsey's experiment at $\pm$ 10 km/sec. This evaluation, remarkable both for the simplicity of its conception and for the elegance of its execution, is thus in entire agreement with the results of the direct measurement of the velocity of light.

The energy levels associated with the rotation and vibration of polyatomic molecules may be investigated both by infrared spectroscopy and by radiofrequency spectroscopy. The former measurements involve a change in the vibrational quantum number of the molecule and, generally, a change in its rotational quantum number. Radiofrequency measurements involve a change of only the rotational quantum number and transitions usually take place in the vibrational ground state. The important point for the measurement of the velocity of light is that infrared spectroscopic measurements are made in terms of wavelength, while radiofrequency measurements are made in terms of frequency. It is possible, using linear polyatomic molecules such as carbon monoxide and hydrogen cyanide, to evaluate the same constant in the expression for the transition energies in terms of both frequency and wavelength, thus giving a value for the velocity of light. The technique has been developed since about 1950 and the most recent value deduced from such measurements is 299,793·7 ± 0·7 km/sec (Rank, Guenther, Saksena, Shearer and Wiggins, 1957). The agreement with direct microwave and optical determinations may be regarded as completely satisfactory and gives valuable additional support for the value for the velocity of light which is now accepted.

VII
Future Measurements of the Velocity of Light

The velocity of light is known to the order of one part in a million. Any attempt to improve the precision of the methods that have so far been used would eventually be limited by the accuracy of the comparison of the appropriate length measurement with the standard of length. Up to 1960 the ultimate standard was the International Prototype Metre, a platinum–iridium alloy bar kept at Sèvres. The instability of this standard and the uncertainty in comparing it with secondary standards limit the accuracy of length determinations based on this standard to about two parts in 10^7. In 1960 the metre was re-defined in terms of the wavelength of an orange line emitted from a discharge in a lamp containing the single isotope krypton-86. This standard of length is reproducible to better than one part in 10^8. Distances are compared interferometrically with the standard; for this purpose the gas discharge laser is likely to be of value. The output from a laser has high spectral purity, so that interference can be observed over very large path differences. The much broader spectral line from a discharge lamp limits the path difference to at most about a metre. Although the laser has not been developed as a wavelength standard, a laser with a wavelength which is known in terms of the Kr^{86} standard could be used in this secondary capacity.

An example of the use of modern techniques is the method proposed by Dr. Froome of the National Physical Laboratory, who was responsible for one of the most precise of the recent measurements of the velocity of light (Froome, 1958). His proposal is similar in principle to his earlier measurement, the determination of the wavelength of microwaves of precisely known frequency, using for the wavelength measurement an interferometric method based on, for example, a microwave version of the Fabry–Perot interferometer. Earlier he used a wavelength of 4 mm; he now proposes to use a

FUTURE MEASUREMENTS OF THE VELOCITY OF LIGHT   37

wavelength of 0·2 mm from a harmonic generator (Froome, 1962) operating at a fundamental wavelength in the region of 9 mm. The klystron which produces this wavelength has a frequency which is precisely known in terms of a primary standard. High accuracy in the measurement of the wavelength of the 0·2 mm waves is achieved by using a large path difference, about 50 m, in the interferometer. The use of very short microwaves not only gives a very large number of wavelengths in this distance, but reduces the effect of diffraction which necessitated in previous measurements a small but significant theoretical correction to the measured value of the wavelength. The path difference in the interferometer is to be measured in terms of the Kr^{86} standard through the use of laser fringes.

Another suggestion for the measurement of the velocity of light is based on the very narrow spectral width of the light from lasers. If a photo-cathode is illuminated simultaneously with the outputs from two lasers operating at slightly different wavelengths the non-linear amplitude response of the photo-cathode produces in the photo-current a component at the difference frequency of the two lasers. Alternatively a single laser producing two different wavelengths may be used. The difference frequency can be measured precisely, and at the same time the wavelength difference between the laser outputs can be measured by observing the fringes formed by the combination of the two outputs of slightly different wavelengths. Bennett and Knutsen (1964) have proposed the use of the 11,522·76 A and the 11,525·02 A lines in the neon spectrum which can simultaneously be produced from a laser. The frequency difference between these lines is about 51,000 Mc/s, which could be generated from the two lines with a suitably designed detection system, and the optical fringes are spaced by 2·9 mm.

If the velocity of light can be determined with an accuracy of better than one part in 10^8 any further increase in the accuracy would require a new definition of the length standard, since the present standard is capable only of defining a length to this accuracy. As an alternative, the behaviour of light could be used to define the standard of length. That is, the metre might be defined as " the distance travelled by electromagnetic radiation in free space in a time equal to $x \times 10^{-9}$ second ", where x would be a number, close to 3·334, which is stated to nine or ten significant figures. Distance measurements would then be made by timing the passage of light or other electromagnetic radiation over the unknown path, in much the same way as the Geodimeter and similar instruments are used at present by assuming a value for the velocity of light.

Fresh proposals for the measurement of the velocity of light appear regularly in the literature. World War II saw the beginning of a period of scientific progress of unparalleled rapidity, and a host of new techniques have been developed for the experimenter. Some of these have yet to be exploited, and others have yet to be discovered. There is no reason to believe that the value of the velocity of light will ever be a closed subject.

References

ANDERSON, W. C. (1937), *Rev. Sci. Inst.* **8**, 239; (1941), *J. Opt. Soc. Am.* **31**, 187.
ARAGO, D. F. J. (1850), *Compt. Rend.* **30**, 489.
ASLAKSON, C. I. (1949), *Nature* **164**, 711; (1951), *Trans. Am. Geophys. Un.* **32**, 813.
BEARDEN, J. A. and THOMSEN, J. S. (1959), *Am. J. Phys.* **27**, 569.
BENNETT, W. R., Jr., and KNUTSEN, J. W., Jr. (1964), *Proc. I.E.E.E.* **52**, 861.
BERGSTRAND, E. (1951), *Arkiv för Fysik* **3**, 479; (1957), *Ann. Franç. Chron.* **11**, 97.
BIRGE, R. T. (1941), *Reps. Prog. Phys.* **8**, 90.
CLELAND, M. R. and JASTRAM, P. S. (1951), *Phys. Rev.* **84**, 271.
CORNU, M. A. (1876), *Annales de l'Observatoire de Paris* **13**, 293.
ESSEN, L. (1950), *Proc. Roy. Soc.* A **204**, 260.
FIZEAU, H. L. (1849), *Compt. Rend.* **29**, 90.
FLORMAN, E. F. (1955), *J. Res. Nat. Bur. Stand.* **54**, 335.
FOUCAULT, J. L. (1850), *Compt. Rend.* **30**, 551; (1862), *ibid.* **55**, 501, 792.
FROOME, K. D. (1952), *Proc. Roy. Soc.* A **213**, 123; (1954), *ibid.* A **223**, 195; (1958), *ibid.* A **247**, 109; *Nature* **181**, 258; (1962), *Nature* **193**, 1169.
GUTTON, C. (1912), *J. de Phys.* **2**, 196.
HOUSTOUN, R. A. (1938), *Nature* **142**, 833; (1941), *Proc. Roy. Soc. Edinburgh* A **61**, 102; (1950), *ibid.* A **63**, 95.
HÜTTEL, A. (1940), *Ann. Phys.* **37**, 365.
KAROLUS, A. and MITTELSTAEDT, O. (1928), *Phys. Z.* **29**, 698.
LUCKEY, D. and WEIL, J. W. (1952), *Phys. Rev.* **85**, 1060.
MACKENZIE, I. C. C. (1954), *Ordnance Survey Professional Papers* No. 19, H.M.S.O., London.
MCKINLEY, D. W. R. (1950), *J. Roy. Astr. Soc. Canada* **44**, 89.
MICHELSON, A. A. (1879), *Nature* **21**, 94, 120, 226; (1882), *Astronomical Papers for the American Ephemeris and Nautical Almanac* **2**, Part 4; (1927), *Astrophys. J.* **65**, 1.
MICHELSON, A. A. and MORLEY, E. W. (1887), *Am. J. Sci.* **34**, 333.
NEWCOMB, S. (1880), *Astronomical Papers for the American Ephemeris and Nautical Almanac* **2**, Part 3; (1885), *Nautical Almanac, Washington*, 112.
PERROTIN, H. (1908), *Annales de l'Observatoire de Nice*, 11.
RANK, D. H., GUENTHER, A. H., SAKSENA, G. D., SHEARER, J. N. and WIGGINS, T. A. (1957), *J. Opt. Soc. Am.* **47**, 686.
ROSA, E. B. and DORSEY, N. E. (1907), *Bull. Bur. Stand.* **3**, 433.

SCHÖLDSTRÖM, K. O. R. (1955), *The Determination of Light Velocity in Öland*, Aga Company, Stockholm.
SILSBEE, F. B. (1955), quoted by Florman (1955).
WALLER, C. K. (1956), *Cartography* 1, No. 3.
WORTHING, A. G. and GEFFNER, J. (1944), *Treatment of Experimental Data*, Wiley, New York.
YOUNG, J. and FORBES, G. (1881), *Proc. Roy. Soc.* 32, 247.

Part 2

Part 2

1
Measurement of the Velocity of Light in a Partial Vacuum*

By A. A. Michelson,† F. G. Pease and F. Pearson

ABSTRACT

The observations were made by the rotating-mirror method, the light passing through a steel tube 1 mile long, evacuated to pressures which ranged from 0·5 to 5·5 mm mercury. By multiple reflections the path length was increased to 8 or 10 miles.

The distance was obtained by reference to a carefully measured base line adjoining the tube.

The time was measured stroboscopically through successive steps by use of a tuning fork synchronized with the rotating mirror, a free swinging pendulum, a chronometer, and wireless signals from Arlington.

There were made 2885·5 determinations of the velocity, the simple mean value of which is 299,774 km/sec, with an average deviation of 11 km/sec from the mean.

INTRODUCTION

The following is a report on the measurements of the velocity of light made at the Irvine Ranch near Santa Ana, California, during the period September 1929, to March 1933. The undertaking was proposed and planned by A. A. Michelson, professor of physics at the University of Chicago and research associate of the Carnegie Institution. Professor Michelson also obtained the funds for the project and lived to see the apparatus installed but was unable to take part in the measurements, which were carried out by F. G. Pease,

---

* *Contributions from the Mount Wilson Observatory, Carnegie Institution of Washington*, No. 522. *Astrophysical Journal*, volume 82, 1935.

† Dr. Michelson died on May 9, 1931, when 36 of the 54 series of 1931 observations had been completed.

of the Mount Wilson Observatory, and F. Pearson, of the University of Chicago.

It will be recalled that a series of measurements of the velocity of light had been made between Mount Wilson and Mount San Antonio in 1924–1926 which gave a value of 299,796 km/sec.* Since the internal agreement of these measures was good, some explanation is desirable as to why it was thought necessary to repeat the experiment at the Irvine Ranch. The measurements involve two distinct elements: first, the time; and, second, the distance. It was estimated that with a rated tuning fork and stroboscopic methods the time of rotation of the mirror could be measured to one part in a million. The time element could therefore be determined with sufficient accuracy.

In the 1924–1926 experiments the determination of the distance required the measurement of a long base line and an extended triangulation from this base into the mountains. Since the Observatory was not prepared to carry out this part of the work, the United States Coast and Geodetic Survey kindly consented to undertake it. Although the splendid work of the Survey in the resulting investigation† had never been excelled, it was felt that the direct measurement of a short base line, without subsequent triangulation, might yield an even higher order of accuracy. Moreover, there was the fact that the results for the velocity of light depend upon the refractive index of the air between the stations, about which little is known except that the total effect is small. The use of a vacuum tube a mile long would eliminate this factor and have the added advantage of giving a small, well-defined image, unaffected by atmospheric disturbances. By allowing the beam of light to traverse the tube eight or ten times, the length of path would be such that the speed of the rotating mirror need not be excessive.

Funds for repeating the investigation with these improvements were generously supplied to the University of Chicago and the Mount Wilson Observatory by the Rockefeller Foundation and the Carnegie Corporation. Through the courtesy of Mr. James Irvine, Jr., a level, unobstructed site on the Irvine Ranch near the city of

---

* *Mt. W. Contr.*, No. 329; *Ap. J.*, 65, 1, 1927.

† A detailed account by William Bowie is given in the *U.S.C. and G.S. Report of Geodetic Observations*, 1923.

Santa Ana, California, was selected for the location of the vacuum tube. Preliminary experiments late in 1929, on a tube 1100 feet long, gave results which were so promising that the construction of the mile-long tube was begun in the winter of 1929–1930.

The United States Coast and Geodetic Survey was again called upon for three successive years to determine the length of the new base line, this time a simple distance between two points about a mile apart. The mean result obtained in these measurements by Commander Garner and Lieutenant Latham, as furnished by the Coast Survey, is 1594·2658 m.

The simple mean of all the readings for the velocity of light is 299,774 km/sec *in vacuo*. Since the values fluctuate somewhat with the time, this mean may differ slightly from what would be obtained if observations were made continuously over an extended period. Series of measures 1–13 and 26–54, made from February 20 to July 14, 1931, gave 299,775 km/sec. Series 14–25, made from March 25 to April 3, 1931, gave 299,746 km/sec. The fact that these mean results differed from each other and from the value 299,796 km/sec obtained on Mount Wilson necessitated additional readings.

Further readings made from March 3 to August 4, 1932, gave a mean value of 299,775 km/sec. If, however, the readings be divided into two groups with an equal number of individual determinations of the velocity, series 55–110 give a value of 299,780 km/sec, while series 111–158 give 299,771 km/sec.

Readings were resumed in December 1932, giving a mean high value of 299,785 km/sec, which dropped to a mean of 299,765 km/sec on January 15 and rose again to the earlier value on February 28. The mean velocity for the 75 series was 299,775 km/sec.

Attempts to explain these variations in velocity as a result of instrumental effects have not thus far been successful.

DESCRIPTION OF APPARATUS

Optical layout.—A diagram of the optical arrangement of the apparatus is shown in Fig. 1.1. Light from an arc lamp A was imaged on the slit C by the condensing lens B. For the first 46 series of the 1931 observations it passed above the right-angle prism I to the upper half of the rotating mirror, D, thence through the plane-parallel glass window L into the tube to the diagonal flat E and the concave mirror *F*. It next passed above the flat mirror *H* and then, by means of repeated reflections at the flat mirrors *G* and *H* until the desired distance had been traversed, formed on the surface of *G* or *H* a magnified conjugate image of the slit *C*. The beam then retraced its path directly below the incoming path and emerged from the tube, striking the lower half of the rotating mirror *D*; it then passed through the reflecting prism *I* on to the crosswires *J* and was observed in the eyepiece *K*.

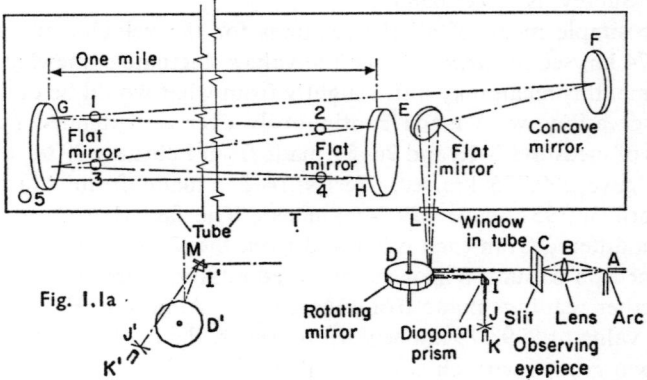

Fig. 1.1. Diagram of optical system.

For series 47–54, to eliminate the effect of any lateral shift of the rotating mirror, the light after passing through the slit entered the 90° prism I' (Fig. 1.1a) and was reflected to the lower half of the rotating mirror at nearly perpendicular incidence, thence into the tube of the flat mirror at an equivalent distance of 4 or 5 miles. The returning beam struck the upper half of the rotating mirror and was then reflected by the silvered surface *M*, which stands directly above the 90° prism, on to the crosswires J^1 and into the eyepiece K^1. Since the advantages of this setup were found to be negligible, the original arrangement of the prism (Fig. 1.1) was used in the 1932 and 1933 measures. It was originally intended to pass parallel light from *F* to *G* and *H*, thence under *G* to a concave mirror about 51 feet beyond it, which would converge the light on to a small concave mirror and thus form a system similar to that used in the Mount Wilson–San Antonio experiments. This auto-collimating system was used in the earlier work because it requires no delicate adjustment, whereas, without the auxiliary mirrors, the flat must be kept accurately aligned.

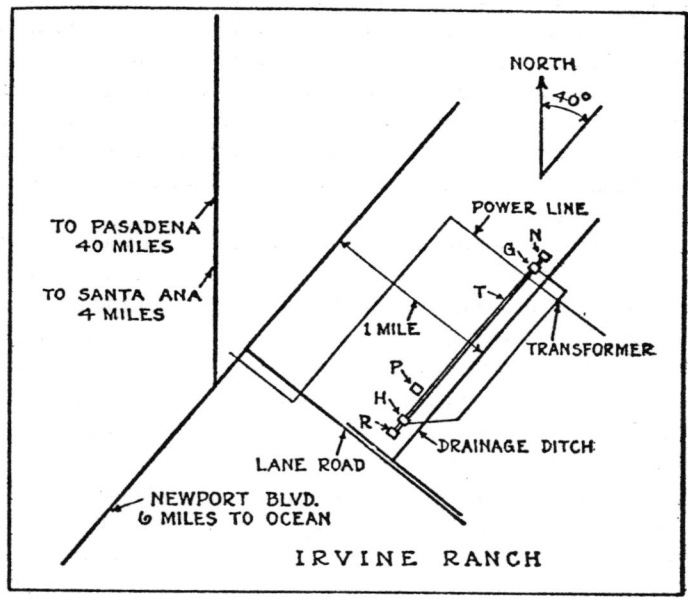

FIG. 1.2. Plan of experiment.

Actual observations showed, however, that the light returned from the conjugate image was brighter than that given by the auto-collimating system; the former arrangement was therefore used throughout the experiments.

The vacuum tube T (Figs. 1.1 and 1.2) is 3 feet in diameter and approximately a mile long. The pipe is interrupted at H and G (Fig. 1.2) by steel tanks which house the 22-inch flats and ended at R and N by tanks containing the concave mirrors.

All operations were conducted from the observing room at H (Fig. 1.2). Here the optical axis was located about 5 feet above the floor, and in the 1931 experiment the slit, condensing lens, rotating mirror, air controls, etc., were mounted outside the tube on a metal table bolted to the cement floor. In the 1932–1933 experiments the slit, prism, rotating mirror, and observing eyepiece were mounted on a heavy cast-iron base, fastened to a solid concrete pier. The pier, together with the metal table and the pendulum case, was fastened to a single massive concrete pier 3 feet thick, whose top lay flush with the floor. The arc lamp, which stood outside the observing room, was surrounded by a metal shield, provided with red-glass window for observing the arc itself. A blackened tube extended from the wall to the condensing lens, and the use of a small aperture, about 0·5 inch square, at the inner end of the tube eliminated much of the undesirable light.

To assist in lining up the mirrors, small 6-volt lamps were inserted at 1, 2, 3, 4, and 5 (Fig. 1.1). Lamps 1 and 3 were placed 100 feet in front of the N 22-inch flat and opposite the centers of the upper and lower halves, respectively, of the mirror. Lamps 2 and 4 were similarly placed in front of the S 22-inch flat. Lamp 5 was opposite the center of the N concave. Lamps 1 and 3 and 2 and 4 were mounted on semaphore arms (electric windshield-wipers) which could be swung into position when needed for aligning. The arc lamp A (Fig. 1.1) was a Mole-Richardson semi-automatic projection lamp, using National cored carbons. It operated on 110 volts and 40 amperes. The arc ran steadily over long periods of time, needing only occasional adjustments by the attendant. The slit C (Fig. 1.1) was adjustable horizontally and vertically and racked for focus. The slit aperture was limited vertically by occulting bars; the slit width was about 0·003 inch.

The rotating mirror D (Fig. 1.1) used in these experiments is of well-annealed optical glass and has 32 faces. It is 0·25 inch thick and 1·5 inches across its diagonals and has a central hole to define its position on the axis. Its angles are correct to $1''\cdot 0$ and its surfaces flat to 0·1 wave. The operating aperture in each direction is 9/64 inch wide and 1/8 inch high. The mounting is one of those used in the Mount Wilson–San Antonio experiments having compressed-air turbine drive capable of rotation in either direction. A slight amount of oil suffices to lubricate the plain journal bearings and the single-ball step-bearing below.

The plane-parallel window L is of crown glass 6 inches in diameter and 0·75 inch thick. At first cemented to the tank flange, it was later held by clamps and atmospheric pressure against a rubber gasket. The diagonal flat mirror E is 5 inches in diameter and 0·5 inch thick. It has motor-driven slow motions about vertical and horizontal axes. The concave mirror F is of glass, 40 inches in diameter and 3·9 inches thick, and has a focal length of 49·28 feet. It is covered with a cardboard screen having an elliptical aperture 12 × 15 inches. The mirror and its mounting were built for the Mount Wilson–San Antonio experiment. The mounting is of cast-iron, in two sections. The lower ribbed base frame rests on three legs, the front one of which is adjustable by motor, which tilts the mirror about a horizontal axis at right angles to the tube axis. The base stands on three cylindrical steel plugs, set in a pier separate from that holding the tank and projecting through three holes in the bottom of the steel tank. Rubber sleeves connecting the tank and the cylinders form a flexible, air-tight joint, thus leaving the mirrors free from any motion the tank itself might have. The upper section of the mounting rests on three balls and has a slow motion in rotation about a vertical axis directly under the face of the mirror. The mirror is held in its cell by three pivoted edge-arcs, the top one having spring contact to prevent excessive loading. Three tangential grooves in the edge of the mirror fit corresponding lugs in the edge-arcs.

The flat mirrors G and H are $22\frac{3}{8}$ inches in diameter and 4 and 5 inches thick, respectively, and are adjustable about horizontal and vertical axes. The details of the mountings are similar to those of the concave mirrors already described. It was found that the silver coatings of the mirrors deteriorated rapidly; Dalton, of the Observatory optical shop, succeeded in coating them with a very thin silver lacquer in such a manner that their optical properties were not impaired. The 90° prism I is blackened on its top surface and silvered and blackened on its diagonal face to prevent any unnecessary illumination in the field of view. For series 1–25 it was mounted directly on the rotating-mirror support, but for the remaining work it was mounted on a shelf attached to the table, thus eliminating displacements due to any possible turbine reaction. Later experiments showed that these displacements were negligible. The micrometer is a simple slide with a single vertical crosswire, moved by a screw of 40 threads per inch, the head of which is divided into 25 parts, making each division equal to 0·001 inch. An eyepiece of 2·5-inch focus was used. A 6-volt lamp with push-button control illuminated the head.

Vacuum pump.—For exhausting the 1100-foot tube, and for preliminary work with the 1-mile tube, a small Kinney rotating-plunger vacuum pump was used. This pump performed so well that in the final work a Kinney rotating-plunger pump of 350 cubic feet of free-air capacity driven by a 15-hp motor was used. These pumps have no internal packing, are oil sealed, and on a closed system produce a vacuum of 0·05 mm. The pump was connected with the

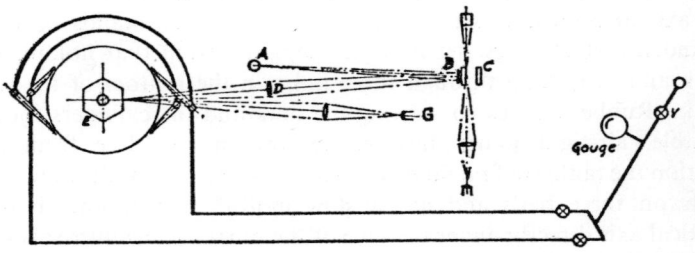

Fig. 1.3. Diagram showing mirror speed control.

tube by a 6-inch steel pipe and was fitted with an automatically controlled check valve to prevent oil from passing into the tube when the pump accidentally stopped. A gate valve, placed at the outer end of the pipe, allowed air to enter the tube when it was necessary to let down the vacuum.

The compressed-air system.—To drive the turbines of the rotating mirror, compressed air at a pressure of 100 lb. was supplied by an ordinary motor-driven compressor. The air was piped to a tank outside the operating room, thence to the valve on the operating table.

The regulating valve V (Fig. 1.3) is an escape valve, consisting of a chamber with an open top fitted with a flat plate with ground surface. A V-socket in the top of the plate holds a $\frac{1}{4}$-inch ball, across which extends a lever held in compression by a spring. The least upward lift by the operator changes the nearly balanced pressures and allows a slight escape of air. After passage through the valve and regulator, the air stream branches, running to the R and L turbines in the rotating mirror.

Electrical system.—Power is obtained from a high-tension A.C. line which crosses the tube about 4000 feet from the S end, where it is stepped down to 110–220 volts and feeds a line paralleling the tube. The power plant located at P (Fig. 1.2) houses the vacuum pump, the air compressor, and the 6-kw motor-generator set supplying current to the arc. All the mirror slow motions are operated by the observer at the micrometer. Those at the S end are handled by direct switches. Those at the N end are worked by a common push-

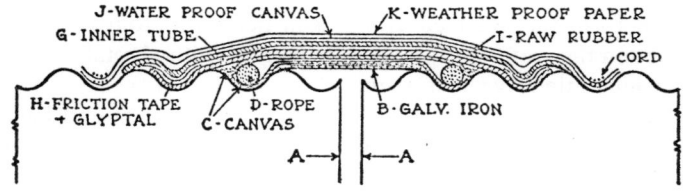

FIG. 1.4. Cross-section of pipe joint.

button line, the proper contact being made through a selector operated by a Selsyn motor.

Pipe.—The pipe is of 14-gauge galvanized Armco steel sheets, 26 inches wide, rolled and corrugated and then formed into sections 60 feet long. The longitudinal seams are double riveted, the circumferential seams single riveted, and all of them soldered. The sections are mounted on wooden trestles and make contact on shoes at the bottom, at 45°, and at the sides.

The pipe sections, A–A, are separated about an inch, each joint being treated as shown in Fig. 1.4. Over each junction was placed a rolled sheet of 12-gauge galvanized iron 8 inches wide, forming a sleeve B whose overlapping ends are bolted to the two ends of the pipes with loosely fitting stove bolts.

Two pieces of canvas C, 11 inches wide, were placed over the sleeve, the inner one tied with a large rope D to fill the groove and the outer one with string. A 32 × 6-inch circular molded inner tube G, previously slit along its inside circumference and stretched over the end of one of the pipe sections away from the canvas, was then drawn over the canvas so that it extended about 2 inches beyond either edge. The edges of the rubber tube were rolled back a few inches and the steel pipe and the inner side of the rubber smeared with rubber cement. The rubber was then rolled flat, smeared for an inch or two with cement, and the joint then wound with several laps of friction tape H and coated with glyptal. In the earlier trials the rubber tube was then directly covered with canvas, but later it was smeared with cement and then covered with a sheet of raw rubber I, 15 inches wide, which becomes tacky with heat and closes any cracks occurring in the inner tube. Waterproof canvas J and a sheet of weatherproof paper K form the outer covering, each tied with stout cord. After completion the seams and joints were painted with glyptal to seal any leaks which might occur in them.

Four manholes give access to the tube, one at each end between the 22-inch flats and the concaves, and two others in the main section of the tube.

Tanks.—The four tanks housing the mirrors and their mountings within the vacuum tube are constructed of $\frac{3}{8}$-inch steel plate, reinforced with appropriate steel sections welded on. The tanks consist of a flat base plate and a removable upper box section, fitting into a groove in the base, which is sealed only with a strip of solder serving as gasket and with Hydroseal. No bolts are used to tie the sections together. The base is a $\frac{5}{16}$-inch flat steel sheet, reinforced with 8-inch I-beams and stands on four small concrete piers. The upper sections are rectangular in shape up to the optical axis and semi-cylindrical above that. Openings in the ends, with flanges 3 feet in diameter and 1 foot long, make the standard flexible connection with the pipe.

The concave mirror tanks R and N (Fig. 1.2) are 4 feet and 4 inches long, 5 feet and $\frac{1}{2}$ inch high, and 4 feet and 6 inches wide, inside measure. The outer ends are fitted with steel drumheads, 3 feet in diameter and 9 inches long, which are joined to the tank with the standard wrapping and supported by a steel frame on two separate piers. Two $\frac{7}{8}$-inch diameter rods about 15 feet long tie these heads to plates sunk about 4 feet underground. Turnbuckles adjust the heads when the 7-ton air pressure is removed from the tube.

The flat mirror tanks H and G (Fig. 1.2) are 6 feet long, 3 feet and 8 inches wide, and 5 feet and 7 inches high inside. A porthole 8 inches in diameter, at a point opposite the face of the 22-inch flat mirror, permits a transfer of measures from the mirror systems inside the tube to the measured-mile piers outside. Tank H has an additional port through which the light passes to and from the rotating mirror. A 4·5-inch i.d. sleeve with a turned outer flange is welded to the tank, the outer end being inclined 10° so that multiple reflections from the window do not interfere with the working beams.

The mirrors and their mountings are those used in the previous velocity experiments, altered to fit the new arrangement and equipped with motor mechanisms for adjustment, all of which are controlled by the observer at the micrometer eyepiece.

SYSTEMS OF MEASUREMENT

Measurement of time.—In velocity-of-light measures made previous to the Mount Wilson–San Antonio experiment the outgoing and return beams were reflected from the same face of the rotating mirror. The return image could always be observed by shifting the eyepiece sideways. In the null method used in the Mount Wilson–San Antonio, and in the Irvine Ranch experiments, the light emerges from one face and is received on some other face—the adjacent face in the Irvine experiment. As the mirror starts rotating, the image gradually passes from the field of view and reappears in the other side of the field only when the rotating mirror is approaching its proper speed.

While several methods are available for measuring the velocity of light with this arrangement, the one chosen is as follows: The rotating mirror is brought into synchronism with a tuning fork whose period of vibration is determined; the position of the image is then read with a micrometer for the right- and left-hand rotations of the mirror. The distance remains fixed. The time interval to be measured is therefore that during which the rotating mirror turns 1/32 revolution, plus or minus a small angle derived from the readings of the micrometer. The period of the fork is determined by stroboscopic methods in terms of free-pendulum beats. Since the period of the tuning fork varies with temperature, comparisons between the fork and the pendulum are made before and after each set of readings. The true time of the pendulum beats was determined before and after each annual series of experiments.

Mirror-speed control.—The light from a 6-volt lamp A (Fig. 1.3), after striking the small mirror B on the tuning fork C, was imaged by the small achromatic lens D on the one polished face of the nut E, which clamps the rotating mirror to its shaft. Since the fork stood vertically, the image vibrated up and down on the nut. The focal length of the lens D was such as to give sufficient amplitude to the motion of the image. Since the nut rotated in a horizontal plane, the image, as the mirror speeded up, passed through a series of vibrating and stationary states and finally reached a permanent stationary state when the beats heard between the fork and the rotating mirror ceased. At this point a second observer made a setting on the return image and a reading of the micrometer. A reversal of the direction of motion of the mirror eliminated any necessity for making a zero reading.

The observer at the eyepiece G (Fig. 1.3) controlled the air supply to the mirror turbines by means of the regulator V already described. The tuning-fork period was adjusted as follows: The light-path was first measured with an ordinary tape. The best available value of the velocity of light divided by thirty-two times this distance gave the whole number of vibrations of the fork per second (N). A fork having a slightly greater period of vibration was selected and filed off until it was very close to the desired pitch. The fork was then mounted in its frame and, while the mirror was rotating in synchronism with it, micrometer readings were made for right and left rotations. This procedure was repeated until the pitch of the fork, as indicated by the small difference between the right and left readings, was a trifle high. The final correction was made by adding a small lump of universal wax to each prong.

Fractional number of beats of the tuning fork.—Light from a filament lamp A (Fig. 1.5) was focused on a narrow slit C and reflected into the pendulum case, whence it was returned by the mirror F on the pendulum and focused by the achromatic lens D on an edge of tuning fork I. When the fork was vibrating, the flashes of light from the pendulum illuminated the fork in various positions and showed to an observer at the telescope J a series of sawtooth images. When the period of the fork was an exact multiple of that of the pendulum, the images as seen against a pointer in the field of view appeared stationary. When the period differed from an exact multiple, the teeth appeared to travel across the field of view. Cognizance was taken of the flashes in one direction only; when the images traveled

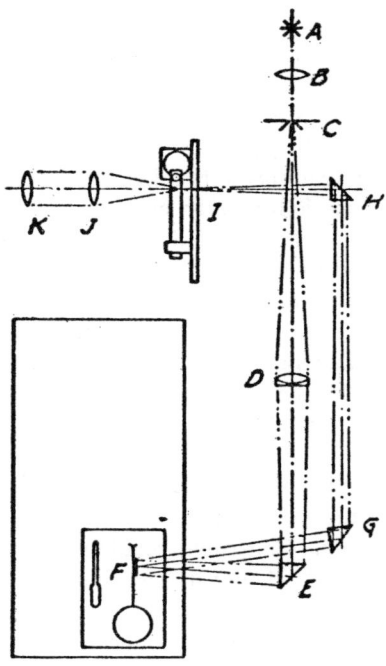

Fig. 1.5. Stroboscopic timing system showing pendulum and fork.

in the same direction as the flash, the fork was losing on the pendulum and the sign of the fractional correction was minus. If the image traveled against the flashes, the sign was plus. If n denotes the number of flashes occurring during the passage from one tooth to the next, the fraction $\pm 1/n$ added to the whole number N gave the period of the fork in terms of the free pendulum.

True period of free pendulum.—The determination of the period of the free pendulum in terms of mean solar time was made in two steps: first, a comparison of beats between the pendulum and a flash box operated each second by a contact-making chronometer; second, a comparison of the chronograph records of second marks from the chronometer with signals (1931 by hand, 1932–1933 by self-recording wireless) recording true time sent by long-wave wireless from Arlington four times each day. This determination was made several times during 1931 and before and after each experiment in 1932–1933, temperatures of the pendulum box and the time intervals being recorded. Since the period of the pendulum varies slightly with the length of swing, readings were taken for several lengths of swing and interpolations made for other swings. The pendulum is one of those formerly used by the United States Coast and Geodetic Survey in the determination of gravity. It is set in motion by an outside lever, swings in a vacuum chamber, and beats half-seconds. Its knife-edges are of agate, rocking on a flat agate plate. When not in operation it rests on auxiliary edges. On the pendulum near its point of suspension and on the shelf which carries it are mounted small speculum-metal mirrors. A dummy bob hangs inside the case carrying a thermometer. The box is of heavy bronze and is provided with adjusting screws and levels. Windows permit observation of the graduated arc and the thermometer and allow the passage of light to and from the mirrors. Consistent readings could not be obtained with the pendulum in 1931, but its inclosure in a constant-temperature case in 1932–1933 eliminated this difficulty. Throughout the experiment the pendulum case was connected with the main vacuum tube.

The flash box used for comparing beats of the pendulum with those of the chronometer consists of a rectangular box and a laboratory telescope mounted on a stand. In the forward end of the box is a 6-volt lamp, a slit, and a shutter, operated by relay with the 6-volt contact circuits of the clocks and controlled by a trigger mechanism which works with great rapidity, giving flashes of very short duration. The images of the slit reflected from the two small mirrors in the pendulum case are seen in the telescope together with a scale on glass placed at the focus. The flash box is usually placed at a given distance from the pendulum box, and the stationary pendulum mirror is adjusted to bring the two images close to one another in the eyepiece field.

The object of the flash box is to determine the times of coincidence between the beats of the pendulum and the chronometer. When the pendulum is started, the chances are that only the fixed-mirror flash will be seen. In a few moments the second image will appear flashing, say in the bottom of the field, and gradually approach the center. When the images are in coincidence, the time indicated by the chronometer is noted. At approximately half the period the moving image appears at the top of the field, and when the down coincidence occurs, the time is noted. The comparisons are continued until coincidences have been obtained over the whole range of temperatures covered by the experiment.

Since the time of vibration depends upon the length of arc of the swing, readings are taken both for the maximum swing and for about one-half the maximum.

If n be the number of seconds between coincidences, then $n/(n \pm 1)$ is the time of one vibration of the pendulum in chronometer seconds, the plus sign being used if the flash travels in a direction opposite to that in which the pendulum swings, and minus if the two travel together. With the apparatus used the pendulum beat faster than the chronometer, and the period of coincidence was roughly 18 minutes.

Two timepieces were used, one a Bond ship's chronometer beating seconds on the relay and missing every fifty-ninth second. The rate was quite constant for a period of 24 hours following its winding. The other was a Constant Frequency Assembly of General Radio make (used late in 1931 only), controlled by an oscillating quartz crystal whose period $50,000\sim$ was reduced through two multivibrators to $1,000\sim$. A unipolar motor of $1,000\sim$ drives a shaft at 10 r.p.s., operating a seconds relay and a synchro-clock. (The rate of this clock was distinctly more constant than that of the chronometer.) The chronograph was a small one of Henson make, driven by a phonograph synchronous motor at a paper speed of 1 inch per second. In 1931 two ink pens were supplied, one operated by the clock circuit beating seconds and the other by hand. In 1932–1933 the records were traced on paraffined paper by self-recording styli.

Time of light-transit.—Let a_1 and a_2 be the right and left readings of the micrometer and r the distance from the mirror to the crosswires. Then the small angle by which the rotating mirror differs in position from 1/32 revolution is

$$\frac{a_1 + a_2}{4r} = \frac{\alpha}{4}.$$

If $1/n$ be the period of the optical beats between the fork and the pendulum and $1/\gamma$ that of the coincidence between the pendulum and the true seconds, and if N be the nearest whole number of the fork, then the correct time elapsed during the passage of the light from the rotating mirror through the tube and back is

$$T = \frac{\left(\frac{2\pi}{32} - \frac{\alpha}{4}\right)(1 - v)}{2\pi(N + n)},$$

which reduces to

$$\frac{\left(1 - \frac{4\alpha}{\pi}\right)(1 - v)}{32N\left(1 + \frac{n}{N}\right)}.$$

Putting

$$\frac{4\alpha}{\pi} = a, \quad \frac{n}{N} = b, \quad v = c,$$

the formula becomes

$$T = \frac{(1 - a)(1 - c)}{32N(1 + b)}.$$

Measurement of distance.—After a conference with Commander C. L. Garner, assistant chief of the Division of Geodesy of the United States Coast and Geodetic Survey, it was decided to lay out the base line about 10 feet to the west of the pipe and place six piers along the line.

Piers E and B (Fig. 1.6) are opposite the flats in tanks G and H, and A and F opposite the concaves in tanks R and N. Piers C and D were intended for use in triangulating into the tube but were not actually used. The mean length of this base line was found by the United States Coast and Geodetic Survey to be 1594·2658 m. The transfer from the base line into the tube and the measurement of the internal distances were made by ourselves in the following manner: Before and after measuring the distance the alignment of the apparatus was carefully checked optically, with the arc on and the rotating mirror stationary, and intersecting marks were drawn on the small diagonal flat at the center of the cone of light. An excellent straight edge (Fig. 1.7) (used to align the tracks of the 50-ft. interferometer on

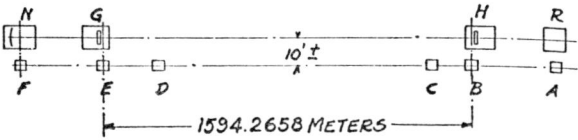

FIG. 1.6. Plan of base-line piers.

Mount Wilson) 12 feet long was passed through the 8-inch diameter opening and supported on frames in a horizontal position. The straight edge was carefully adjusted parallel to the face of the flat by calipering at a, b, and c, and the separation a noted. A mercury plumb bob was then dropped from its edge and its position marked on the bronze plate in the pier. The bob was rotated to determine its neutral position and allowance made for the thickness of the string. If d be the distance of the bob from the bench mark, the simple summation $d \pm a$ gave the position of the south flat with respect to pier B and of the north flat with respect to pier E. The distance between the top of the S 22-inch flat and the center of the S concave was measured with a steel tape under 16·7-lb. tension, to which a steel scale was clamped to make contact with the concave. Temperatures were recorded and the mean of three readings taken as a correct one. This inclined distance was then corrected to give the horizontal distance between the mirrors. The distance from the S concave to the small diagonal flat was measured with the same tape, tension, and scale. The addition of a trammel bar clamped to the tape at the end near the flat allowed its point to be adjusted to a center mark on the flat.

The distance from the small diagonal flat to the rotating mirror was taken through the window-opening by means of a trammel bar and measured on the tape. Corrections were then made for the path

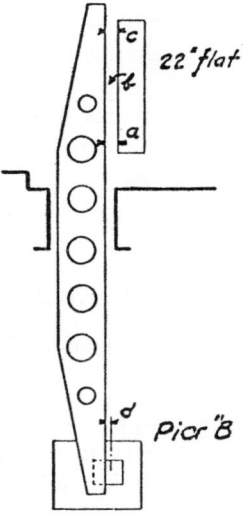

FIG. 1.7. Diagram showing method of transferring mirror position to base-line pier.

length through the glass 19·33 mm thick, at an inclination of 10° ($n_d = 1·52$) equivalent to a vacuum thickness of 29·39 mm. The air path outside the window was reduced to vacuum thickness by using 1·000295 as the index for air.

Table I gives the various results for the length of the base line as measured by the United States Coast and Geodetic Survey. The data show a slight progressive increase in the means of each year's measures, which may be considered as a real change in the earth's surface. Interpolated values of the distance might have been used for

TABLE I

Length of Base Line, Piers B–E

Date	Observer	No. of Traverses	Distance in Millimeters
U.S. Coast and Geodetic Survey			
1931 Feb. 27—Mar. 2	Garner	9	1594259·2
1932 Jan. 11—Jan. 14	Latham	8	1594265·8
1933 Jan. 14—Feb. 17	Latham	31	1594272·3
Mean			1594265·8
Mount Wilson Observatory			
1933 July 17–20	Pease	8	1594263·8

TABLE II

Length of Light-path in Millimeters

Path	1931	1932	1932–1933
Rotating mirror to 5-in. diagonal corrected for air path	1661·3	1662·9	1661·9
5-in. diagonal to S concave	13092·7	13098·6	13102·2
S concave to S 22-in. flat	13714·1	13713·5	13716·8
D_1 = sum of above three	28468·1	28475·0	28480·9
S 22-in. flat to pier B	0·6	2·0	0·6
Pier B–E (mean)	1594265·8	1594265·8	1594265·8
Pier E–N 22-in. flat	14·4	14·4	16·3
D_2 = sum of above three	1594280·8	1594282·2	1594282·7
$D_{10} = 2D_1 + 10D_2$	15999744·2		
$D_8 = 2D_1 + 8D_2$	12811182·6	12811207·6	12811223·4

the computations of the velocity of light, but, owing to large variations in the velocity results, it was thought that the simple mean value could be used without prejudice to the results. An earthquake which occurred on March 10, 1933, may explain the reduced value of the base line measured by Pease in July, 1933. Table II gives the various lengths involved in the light-path.

CALCULATION OF VELOCITY

The formula for calculating the velocity of light from the observed data is

$$V = \frac{D}{T} = \frac{32ND(1 + b)}{(1 - c)(1 - a)},$$

or, with sufficient approximation, since b and c are small,

$$V = 32ND(1 + a + b + c).$$

The presence of a slight residuum of air in the tube at temperature T necessitates a small correction e, whence the formula becomes

$$V = 32ND(1 + a + b + c + e) = 32NDf.$$

The value of e is computed from the formula for the refractive index of air given in the *Smithsonian Physical Tables*:

$$n - 1 = \frac{0 \cdot 0002931 \times P' \times 1333 \cdot 2}{1 + 0 \cdot 00367 t \times 1 \cdot 0136 \times 10^6},$$

where P' is the pressure in millimeters of mercury and t the temperature in degrees centigrade. For the 10-mile distance a tuning fork giving 585 v.p.s. was used. For the 8-mile distance a fork was used, for which the whole number of vibrations was 365 or 366, N thus being either 2×365 or 2×366. The various values of $32ND$ used in the experiment are shown in Table III.

TABLE III
Values of $32ND$

Velocity	Vibrations per Second	1931	1932	1932–1933
V_{10}	585	299515·2		
V_8	365×2	299269·2	299269·8	299270·2
V_8	366×2	300089·1	300089·7	

METHOD OF OBSERVATION

With the pump running continuously, the vacuum remained approximately constant, the pressure ranging from 0·5 to 5·5 mm according to the amount of leakage. During the day the influence of the sun on the air in the tube was such as to distort and blend the several images corresponding to the various distances and thus prevent work at such times. After sundown the images improved rapidly and in a half-hour were easily separable from each other. The tuning fork was started in advance in order to give it a chance to warm up. The procedure in lining up the mirrors was as follows: With the arc lamp on, the outgoing beam from the rotating mirror was centered on the upper half of the S concave mirror by means of the 5-inch diagonal flat. The arc lamp was then cut off; lamp 1 was turned on and the concave mirror in the S end adjusted until the light appeared in the center of the field. The N 22-inch flat was then adjusted until the image of the lamp as seen in the mirror G appeared superposed on the light itself. Next, lamp 2 was turned on, and a more careful adjustment of the N 22-inch plane was made in order to bring lamps 1 and 2 into superposition. Lamp 1 was then turned off, and the image of lamp 2 reflected from the S 22-inch flat was superposed on itself by adjusting the flat. Light 3 was next turned on and the S mirror again carefully adjusted to superpose lamps 2 and 3. When this had been done, lamps 3, 4, and 5 were all superposed. A number of images other than the one desired appeared in the field of view; but a study of their positions and foci soon showed which one was wanted.

This method of alignment was evolved when use of the N concave mirror was contemplated. It was found convenient to use the same method when adjusting for work without this concave, save that a small alteration in the inclination of the S 22-inch flat slightly raised the beam so that at the last reflection it fell with normal incidence on one or the other of the 22-inch flats. During all these operations the images were observed through the widened slit, which was then narrowed and placed in the observed focus of the light corresponding to the distance at which one wished to work. Since the field was divided, images also appeared in the eyepiece field. A slight adjustment of the 5-inch flat made the images in the eyepiece and behind the slit of equal intensity. The position of the image in the slit was carefully defined by occulting bars, and care was taken to line up the arc on the axis thus defined. When finally the arc was again turned on, the return light was visible in the eyepiece.

When the rotating mirror was set in motion, images corresponding to each of these reflections (because of diffusion and spreading of the beam) were seen in the field at distances from the crosswire proportional to the distance the light had traveled. Experience enabled one to make slight adjustments which concentrated the light in the image which was to be used. The apparatus was then ready for measurements. Observer A made a reading for the stroboscopic comparison of the fork and the pendulum (n), then read the temperature of the pendulum case (t). The mirror was then brought up to speed by manipulating the air control until the stroboscopic image was stationary. Observer B set the crosswire on the image and recorded the reading. In series 1–25 of 1931, 10 such readings were taken before the direction of rotation was reversed. To minimize the effect of a slight drift which had been noticed, subsequent readings (series 26–46, 1931) were made, 5 in the first direction, 10 in the opposite, and finally 5 more in the first direction. In series 47–54, 1931, 5 alternate sets of 5 readings were taken; the mean of 1 and 3 was used against 2, 2 and 4 against 3, and 3 and 5 against 4. The first readings of the successive group were alternately L and R. This method was used in all the remaining work except a few cases in which 7 sets were made. Observer A repeated the readings for n and t between each two sets, and the mean for the beginning and end of each set was used for the set.

Since the total angle measured was exceedingly small, the lever arm r was measured with an ordinary rule. Allowance was made for the prism thickness except for series 47–54 (1931), for which a mirror was used. Even with the vacuum as low as 0·5 mm, sufficient air remained in the tube to be affected by temperature conditions. The best images were obtained when a quiet fog settled around the tube, evidently providing a constant temperature throughout the tube. On days when the sun shone, the images drifted completely out of the field, usually returning again at night. A tube wet with dew gave comatic images when the wind blew. In 1931 it was noted that the interval during which work could be done became less and less as summer advanced, owing to the fading of the image and the necessity for constant readjustment. A satisfactory explanation for fading may be found in temperature deformations of the 22-inch flat glass mirrors which dispersed the light. In 1932 and 1933 it became the practice to work principally during the early hours of the night, since

TABLE IV

Sample Set: Observations and Reductions, June 30, 1932

(1) L	(2) R	(3) L	(4) R	(5) L	(6)
13·6	19·2	14·0	18·8	13·2	$N = 365 \times 2$
13·8	19·6	14·2	18·9	14·2	B 7.24 P.M.
13·7	20·0	14·0	19·2	13·5	E 7.30
13·4	19·9	13·1	19·5	13·6	p 3·4 mm
13·9	20·0	13·5	19·4	14·2	n_b 2/26 0·07692
					n_e 3/35 0·08571
					0·16263
68·4	98·7	68·8	95·8	68·7	$n + 0·08132$
					t_b 29·7°, t_e 29·7°, t 29·7°C
13·68*	19·74	13·76	19·16	13·74	s_b 4·1, s_e 4·1, s 4·1 mm
13·76	19·16	13·74			r 11·80 in.
					d − 0·00571 in.
13·72	19·45	13·75			a − 0·0006161
19·74	13·76	19·16			b + 0·0002228
					c + 0·0008314
−6·02	−5·69	−5·41			e + 0·0000012
−5·69					f 1·0016715
−5·41					V 299769·81 km/sec.
−17·12					
− 5·71					

* Mean of five readings above.

otherwise much valuable time during the interval of good observing conditions would be lost in resetting on the following night. A few observations were, however, made around midnight and 3.00 A.M.

Typical observation.—Table IV shows a typical set of observations, together with the values of the various factors calculated from the data. Columns 1, 3, and 5 give the micrometer readings made for a left-hand, columns 2 and 4 for a right-hand, rotation of the mirror, the unit being 1/1000 of an inch. Immediately below are their sums and mean values. The mean of the differences,

$$\frac{L_1 + L_3}{2} - R_2, \quad \frac{R_2 + R_4}{2} - L_3, \quad \frac{L_3 + L_5}{2} - R_4,$$

shown farther below, gives the value of d in column 6 which is used to calculate $\alpha = d/r$. Column 6 also lists the time of starting and various other data as follows: p is the pressure in the tube; n_b and n_e the observed and n the mean value of the fractional number of vibrations of the fork; t_b and t_e the observed and t the mean temperature of the pendulum case; s_b and s_e the observed and s the mean swing of the pendulum; r is the measured distance from the rotating mirror to the crosswires, with corrections added for glass thickness when the prism is used. Values of $a = 4\alpha/\pi$, $b = n/N = n/365 \times 2$ are calculated from the data. The value of $c = v$ is taken from the chart showing the true time of the pendulum beat for the mean temperature t and the swing s. The residual air correction e is taken from the chart showing the relation between n, p, and the tube temperature t_t; f is the algebraic sum, $1 + a + b + c + e$, by which $32ND$ is multiplied to give the velocity V.

OBSERVATIONS

Table V shows two typical series of observations on successive nights. Column 1 indicates the number of the series, column 2 the date and hour of observation, column 3 the itemized values of the deflections found by subtracting the midvalue of the micrometer reading from the mean of those each side of it, and column 4 the mean value of the deflection. Columns 5, 6, 7, 8, and 9 give the values of a, b, c, e, and f. Column 10 gives the resulting velocity of the set of three velocity determinations, column 11 the residuals in km/sec for the velocity relative to the mean for the group, and column 12 the number of single determinations of the velocity.

Table VI shows the results from each of the 233 series of observations. Details are as follows:

Hours of observation (*col.* 3).—Many of the records of the 1931 observations are incomplete as to the time of beginning and ending

TABLE V
Two Typical Series of Observations

Series	Date, P.S.T.	d		a	b	c	e	f	V	v	Wt.
		Individual	Mean								
(1)	(2)	(3)	(4)	(5)	(6)	(7)	(8)	(9)	(10)	(11)	(12)
98	1932 Apr. 28										
	h m										
	8 25	421,417,392	−0·00410	+0·0003934	+0·0004475	+0·0008327	+0·0000014	1·00016750	299,771	−14	3
	8 46	340,377,391	369	3541	4852	8340	14	16747	771	−14	3
	9 5	450,445,446	447	4289	5125	8347	14	17775	802	+17	3
	9 16	357,382,400	380	3646	5518	8350	14	17528	794	+ 9	3
	9 24	318,313,301	−0·00311	+0·0002984	+0·0005819	+0·0008354	+0·0000014	1·00017171	299,785	+ 0	3
	Mean								299,785	±11	15
99	1932 Apr. 29										
	h m										
	8 31	704,688,697	−0·00696	+0·0006103	+0·0003086	+0·0008340	+0·0000017	1·017546	299,795	+16	3
	8 40	669,648,628	648	5682	3255	8347	17	17301	788	+ 9	3
	8 50	591,541,510	550	4823	3479	8354	17	16673	769	−10	3
	9 5	539,542,523	535	4584	3923	8357	17	16881	775	− 4	3
	9 18	505,518	511	4378	4012	8357	17	16764	772	− 7	2
	9 31	558,521,513	−0·00531	+0·0004549	+0·0004065	+0·0008363	+0·0000017	1·0016994	299,778	− 1	3
	Mean								299,779	± 8	17

TABLE VI—*Observations*

(1) Series	(2) Date	(3) P.S.T.	(4) Mean Velocity	(5) Residuals	(6) A.D.	(7) Wt.
	1931	h m h m	km/sec.		±	
1	Feb. 19	9 25—10 5	299,792	+8, −7	7	2
2	20–21	— 2 0	774	+26, −27, −45, +4, +27, +15	24	6
3	23	9 0—10 20	767	+27, +7, −34	23	3
4	24	9 50—11 10	760	−1, +18, −17	12	3
5	26	12 30— 2 40	773	−15, +37, −24, −8, +10	19	5
6	26	8 20—10 0	768	+40, +10, −15, −17, −18	20	5
7	26	10 1—11 10	758	−8, +17, +1, 0, −10	7	5
8	26–27	11 10— 2 2	766	−1, −11, −30, +29, +9, +10, −6	14	7
9	27	8 16— 9 26	796	+66, +33, +33, −23, −38, −12, −35, −25	33	8
10	27	9 27—11 5	772	+26, −9, +2, −12, −13, +5, +9, −8	10	8
11	Mar. 1	9 0—11 10	795	−2, +18, −8, −2, −6	7	5
12	2	8 22— 9 4	756	−28, −29, +49, −23, +31	32	5
13	2	9 5— 9 46	772	+23, +9, −22, −3, −7	13	5
14	25		750	+14, +19, −17, −21, −18, +22	18	6
15	26		750	−21, −3, +25, −4, +2	11	5
16	26		745	−28, −9, +20, +7, −10, +10	10	7
17	27	8 33— 9 31	748	−10, +6, −16, +9, +6, +6	9	6
18	27	9 35—11 9	743	+10, −3, −11, −3, −11, +18	8	6
19	30	9 0— 9 59	744	−14, +18, +43, −31, −13, 0, −3	17	7
20	30	10 0—11 12	745	−17, +5, −3, +13, +24, −12, −11	12	7
21	31	1 20— 1 43	728	−2, +4, −24, +22	13	4
22	31	8 30— 9 24	755	+10, +36, −33, +6, −24, +6	19	6
23	31	9 25—10 0	757	0, +11, 0, −17, +7	7	5
24	Apr. 3	9 55—10 34	741	+14, +1, −1, −11, −3	6	5
25	3	10 36—11 18	741	+1, +8, −14, +16, −11	10	5
26	8	8 25— 9 0	755	−52, +38, +14	35	6
27	15	8 0— 9 28	754	+5, −63, +31, +11, +15	25	10
28	15	10 14—10 49	770	+7, +15, −22	15	6
29	16	7 50— 9 12	775	+8, −13, −27, +10, +11, +13, +1, +2	11	16
30	16	9 17—10 31	776	−2, −12, −7, −4, +16, −1, −6, +16	8	16
31	17	8 23— 9 18	771	+4, +12, −7, −7, −2	6	10
32	17	9 19— 9 51	776	−7, −11, +18, 0	9	8
33	21	7 45— 8 38	793	+37, +11, −12, −6, −13, +1, −18	14	14
34	21	8 39— 9 39	779	−20, +10, −6, +8, +4, +4	9	12
35	21	9 41—10 33	772	+5, −14, +1, −3, +8, +3	6	12
36	22	7 54— 8 22	778	+6, +5, +1, −12	6	8
37	May 7	9 35—10 24	789	−13, +14, +12, −13	13	8
38	14	7 40— 8 30	774	−7, +1, +3, +2, +6, −6	4	12
39	14	8 31— 9 41	770	+8, −11, −2, −7, −3, −8	6	12
40	14	9 42—10 30	777	−7, −17, +6, +15, +3	10	10
41	14–15	11 48—12 36	773	0, −3, −1, 0, +5, −2	2	12
42	15	12 37— 1 15	775	+1, −8, −7, +3, +11	6	10
43	15	7 29— 8 23	779	+15, +4, −9, +9, −19, −11, +11	11	14
44	15	8 52— 9 39	767	+10, −4, −9, +9, +7, −7, −6	7	14
45	15	9 56—10 38	299,775	−6, −4, −3, +15, +12, −16	9	12

TABLE VI—continued

(1) Series	(2) Date	(3) P.S.T.	(4) Mean Velocity	(5) Residuals	(6) A.D.	(7) Wt.
	1931	h m h m	km/sec.		±	
46	May 16	12 5—12 58	299,778	$-1, -9, -12, +1, +6, +2, +10, +2$	5	16
47	July 1	8 8— 9 52	776	$-10, +12, -18, -5, -9, +30$	13	16
48	3	7 15— 8 18	775	$+6, 0, -7, +14, -13$	7	13
49	6	7 33— 8 22	774	$+8, +1, -8$	6	9
50	7	7 15— 8 3	773	$-23, -25, -2, +25, +25$	20	15
51	7	8 14— 9 18	764	$-12, +26, +2, -7, -10$	11	15
52	8	7 15— 8 45	777	$+12, +13, +6, -16, -6, -9$	7	18
53	13	7 6— 8 25	779	$-30, +21, +5, +15, +30, -41$	24	18
54	14	7 25— 8 24	778	$+20, -6, -10, +9, -14$	12	15
	1932					
55	Mar. 3	9 20— 9 49	815	$-2, +1$	1	6
56	4	7 23— 8 11	772	$-2, -2, -1, +7$	3	12
57	4	8 20—10 1	814	$+16, -37, +16, +8, -15, -6$	17	14·5
58	4	10 23—11 14	815	$-4, +1, +12, -7$	6	12
59	8	7 55— 8 45	796	$-3, -6, +5, +4$	4	12
60	8	8 49— 9 54	782	$+1, +6, -18, +12$	9	12
61	8	9 56—10 59	789	$-5, +16, +24, -19$	14	10
62	9	7 39— 8 40	800	$-4, +11, +11, -1, -14$	8	15
63	9	9 6— 9 52	821	$+25, -12, -1, -10$	12	12
64	9	10 15—11 11	809	$-12, -1, +11, -6, +9$	8	15
65	10	7 35— 9 39	751	$+14, +14, +7, -8, -16, +5$	11	13
66	11	7 58— 8 40	789	$+23, -11, +4, -15$	12	12
67	11	8 49—10 11	766	$+1, +10, +11, -2$	6	12
68	11	10 12—10 57	772	$-17, +10, +5, +1$	8	12
69	15	7 59— 9 34	773	$+11, -7, +10, -16$	11	12
70	15	10 13—10 41	774	$-5, -15, +20$	13	9
71	16	7 51— 8 27	775	$-2, -7, 0, +11$	5	12
72	16	8 30— 9 25	779	$+5, -5, -6, +6$	5	12
73	16	9 50—10 42	784	$+2, +10, -9, -18, +9$	10	13
74	16	11 3—11 46	764	$-20, +4, +15, +4$	11	12
75	17	7 7— 8 3	794	$+58, +1, -17, -31, -11$	24	15
76	17	8 6— 9 37	776	$+16, +23, -30, -1, -10$	16	15
77	18	7 35— 8 53	743	$-19, -37, -7, 0, +21, +43$	21	18
78	18	9 10— 9 50	792	$+9, -11, +3$	8	9
79	Apr. 7	9 28— 9 36	736			3
80	8	9 50—10 40	787	$-42, +31, +11$	28	9
81	12	7 26— 8 44	770	$+11, +9, +12, -10, -13, -11$	11	18
82	12	9 31—10 13	779	$+9, 0, -12, +4$	6	12
83	13	7 35— 8 22	780	$-18, +30, -1, -2, -9$	12	15
84	13	8 25— 9 02	783	$-6, -2, +9$	6	9
85	14	7 25— 8 27	788	$0, -7, +6, +29, -32$	15	14
86	14	8 36— 9 22	768	$+4, +7, -10, -1$	5	12
87	15	7 23— 8 11	768	$-57, -51, +17, +36, +39$	38	14
88	15	8 27— 9 8	807	$+16, -1, +17, -16, -15$	13	15
89	18	8 32—10 9	786	$-22, +20, -12, +13, -1$	8	25
90	20	8 2— 8 30	299,789	$0, 0$	0	10

Table VI—continued

(1) Series	(2) Date	(3) P.S.T.	(4) Mean Velocity	(5) Residuals	(6) A.D.	(7) Wt.
	1932	h m h m	km/sec.		±	
91	Apr. 21	7 14— 8 18	299,773	+7, +45, −28, −23	15	20
92	21	8 23— 9 26	761	−24, +21, +1	8	15
93	26	7 34— 8 15	776	−29, +19, +7, +6	15	12
94	26	8 43— 9 26	783	−16, +9, +5, 0	7	12
95	27	7 50— 8 55	759	−23, −27, +23, +14, +16, −2	17	18
96	27	8 59—10 16	790	−35, +5, +5, +5, +13, +5	11	18
97	28	7 34— 8 23	784	+21, +6, +3, −25, −3	12	15
98	28	8 25— 9 33	785	−14, −14, +17, +9, 0	11	15
99	29	8 31— 9 40	779	+16, +9, −10, −4, −7, −1	8	17
100	May 3	7 44— 8 25	766	+3, +7, +10, −21	10	12
101	3	8 41— 9 58	776	+6, −11, +5, +16, −16	11	15
102	4	7 39— 8 33	787	+1, −3, +14, −2 −10	6	15
103	4	8 34— 9 49	780	+6, −12, +3, −3, +6, −2	5	18
104	5	7 31— 8 03	770	+10, −5, −6	7	9
105	6	7 34— 8 16	779	0, +8, −9, 0	4	12
106	10	7 37— 8 57	770	+7, −10, +4	7	9
107	11	7 40— 9 1	786	+1, +25, +1, −14, −8, −8	9	18
108	12	7 36— 8 25	767	−6, +2, −6, +2, +8	5	15
109	12	8 47— 9 34	775	−7, +1, +2, −3, +7	4	15
110	13	7 38— 8 38	765	−4, −11, +19, +13, −8, +1, −13	10	21
111	13	9 0— 9 18	770	−4, +4	4	6
112	17	7 40— 9 9	773	+9, −10, +4, −3	6	12
113	18	7 36— 9 5	775	−42, +21, +14, +7, +23, +26, +7, +11	19	20
114	19	7 45— 9 7	781	+6, −9, +16, −8, 0, +2, −8	7	21
115	20	7 32— 8 42	759	−7, +50, −12, −4, −9, +5	10	11
116	24	7 39— 8 41	774	+9, +4, 0, −11, −2	5	15
117	26	7 27— 8 23	765	+8, −15, −11, −7, +15	11	13
118	27	7 34— 7 42	760			3
119	June 1	8 5— 9 13	778	−28, +4, +20, +13, −10	15	15
120	2	7 34— 8 47	787	−11, −3, +6, +2, −1, +12, −5	6	21
121	3	7 24— 8 14	780	+7, +12, +6, −6, −18	10	15
122	3	8 16— 8 58	775	−2, −17, +1, +18	10	12
123	6	7 38— 8 55	759	−6, +8, −8, +2	7	8
124	7	7 32— 8 56	758	−11, −7, −5, +1, +10, +12	8	18
125	8	12 12—12 21	781			3
126	9	7 32— 8 52	762	−13, −5, −2, −8, +4, +17, +6	8	21
127	10	7 36— 8 45	757	−11, −11, +12, +7, −9, +8, +5	9	21
128	13	7 31— 8 52	782	−13, +4, +14, +4, +3, −6, −5	7	21
129	14	7 24— 8 41	781	0, −4, +19, +23, −14, −7, −18	12	21
130	15	7 24— 9 11	774	−1, +15, −2, +2, −7, −3, −4	5	21
131	16	7 27— 8 38	772	−23, +11, +13, +13, −17, +4, 0	12	21
132	17	7 46— 8 32	763	−7, +10, +1, −3, −1	4	15
133	20	7 46— 8 24	735	+4, −11, +45	12	6·5
134	21	7 44— 8 43	762	−2, −6, −12, 0, +9, +11	8	18
135	22	7 35— 8 54	299,762	−2, +12, −4, −20	7	10

Table VI—continued

(1) Series	(2) Date	(3) P.S.T.	(4) Mean Velocity	(5) Residuals	(6) A.D.	(7) Wt.
	1932	h m h m	km/sec.		±	
136	June 23	7 23— 8 33	299,772	+11, −3, +9, −21, +1, +1, +2	7	21
137	24	7 21— 8 30	770	+16, −8, −4, −5, +4, +1, −5	6	21
138	27	7 28— 8 35	785	+3, −25, +11, −8, −8, +18, +3	11	20
139	28	7 25— 8 27	773	+7, −2, −10, +5, −15, +12, +3	8	21
140	29	7 45— 8 47	770	−20, −14, +6, +2, +12, +21, −8	12	21
141	30	7 15— 8 19	786	−4, −16, +21, +14, 0, −11, −5	10	21
142	July 1	7 50— 8 35	776	+8, −11, +11, +1, −10	8	15
143	5	7 28— 8 34	768	−11, −1, −1, +19, +3, +12, −19	9	20
144	6	7 22— 8 24	757	−10, +4, −2, +8, +13, +1, −14	7	21
145	7	7 37— 8 15	775	+1, +3, +1, −5	2	12
146	8	7 24— 8 21	773	−9, +6, +11, +4, −6, −5	7	18
147	11	7 23— 8 11	775	−5, +5, −7, −1, +9	5	15
148	12	7 56— 8 42	768	−16, −2, +25, −2, −5	10	15
149	13	7 28— 8 23	778	−8, −12, +1, +25, −4, 0	9	16
150	14	7 27— 8 41	781	−23, +12, 0, +21, −10	13	15
151	15	7 22— 8 37	770	−28, −57, +22, +13, +16, −3, −4	14	15·5
152	18	8 2— 8 43	776	+5, −9, −16, +6	7	9
153	20	8 1— 8 43	760	−5, +4, +11, −5, −6	6	15
154	21	7 28— 8 13	783	+1, +1, +10, −9, −3	5	15
155	22	7 27— 8 15	765	−6, −9, +12, +10, −7	8	15
156	25	8 7— 8 43	777	−8, +1, +20	7	7
157	Aug. 3	7 38— 8 33	752	−9, −20, +24, +21, −17	18	15
158	4	8 7— 8 48	779	−4, −13, +18, −2	8	10
159	Dec. 3	8 5— 9 10	805	−2, +1, +1	1	9
160	5	8 36— 9 30	756	+2, −2, 0	1	9
161	5–6	11 55—12 51	776	−26, −12, +22, +12, +3	15	15
162	6	3 22— 4 9	784	+8, +3, +15, −26	13	12
163	6	7 15— 8 22	795	+19, +6, +2, −13, −15	11	15
164	7	7 19— 8 7	792	+6, −1, +4, −19, +9	8	15
165	7–8	11 54—12 40	784	+9, +4, +1, −8, −6	6	15
166	8	3 6— 3 52	772	+12, −8, −13, +6, +2	8	15
167	9	12 3—12 50	786	+11, +16, +1, −5, −23	11	15
168	9	3 9— 4 5	765	−13, +6, +16, −8, −1	9	15
169	9	7 31— 8 21	791	+1, +8, +8, −1, −17	7	15
170	10	12 25— 1 14	797	+43, +5, −27, −25, +3	21	15
171	19	8 4— 8 45	775	−17, −18, +14, +22	18	12
172	20	7 38— 8 31	770	+2, +21, −2, −10, −12	9	15
173	20–21	11 57—12 44	764	−11, −3, +6, +9	7	12
174	22	7 14— 8 4	779	−9, −21, −9, +16, +24	16	15
175	23	12 6— 1 0	777	−3, +3, −9, +5, +5	5	15
176	23	7 43— 8 34	776	+11, +14, −15, −14, +3	11	14
177	24	12 7—12 54	755	+29, −15, −2, −13	15	12
178	24	3 19— 3 56	760	+16, −19, +21, −18	18	12
179	27	8 30— 9 15	774	−7, −14, +24, −3	12	12
180	27–28	11 51—12 38	299,777	−13, +13, +4, −3	8	12

THE VELOCITY OF LIGHT

TABLE VI—*continued*

(1) Series	(2) Date	(3) P.S.T.	(4) Mean Velocity	(5) Residuals	(6) A.D.	(7) Wt.
	1932	h m h m	km/sec.		±	
181	Dec. 28	2 53— 3 42	299,764	+9, +2, −15, +4	8	12
182	28	7 36— 8 21	776	+33, −47, +18, −4	26	12
183	28–29	11 46—12 39	777	−10, −12, +5, +17	11	12
184	29–30	11 52—12 26	782	+15, +9, 0, −23	12	12
185	30	8 29— 9 05	774	−29, +*11*, 0, +17	15	11
186	31	12 11—12 30	780	−35, +35	35	6
	1933					
187	Jan. 3	7 30— 8 4	775	−5, +9, −3, −2	5	12
188	5	7 27— 8 10	765	+4, −13, +13, −4	8	12
189	6	12 0—12 32	796	+3, +2, −26, +21	13	12
190	9	8 3— 8 28	763	+9, −9, 0	6	9
191	10	12 5—12 32	770	+1, +6, −7	5	9
192	10	7 29— 8 2	764	+29, −6, −15, −8	14	12
193	11	7 17— 7 48	758	+39, −10, +3, −32	21	12
194	12	12 2—12 48	764	+*6*, +9, +5, −20	11	10
195	12	7 6— 7 40	753	−3, −23, +8, +18	13	12
196	12–13	11 54—12 31	766	+20, −4, −21, +5	12	12
197	13	2 56— 3 30	765	+21, −9, −5, −7	10	12
198	13	7 10— 7 30	786	−5, +5	5	4
199	16	7 57— 8 30	761	+10, +8, −5, −13	9	12
200	16–17	11 53—12 24	749	+7, +8, −6, −9	8	12
201	17	3 3— 3 37	767	0, −7, +2, +5	4	12
202	17	7 57— 8 25	764	−19, +7, +12	13	9
203	18	12 0—12 31	764	−12, −11, +23	15	9
204	18	7 25— 7 56	777	−1, −18, +9, +9	9	12
205	19	12 13—12 45	774	+6, +10, +5, −21	10	12
206	19	2 59— 3 34	771	−2, −1, +10, −7	5	12
207	19	8 18— 8 42	759	+17, −15, −2	11	9
208	25	7 48— 8 28	771	−10, −4, +2, +12	7	12
209	26	12 16—12 41	769	−19, +20, 0	13	9
210	26	3 12— 3 42	770	+2, −22, +19	14	9
211	26	7 19— 7 59	765	+4, −14, −4, +14	9	12
212	27	7 28— 7 59	779	+4, −2, +4, −6	4	12
213	31	7 9— 7 42	771	+3, −4, +7, −7	5	12
214	Feb. 1	12 4—12 28	774	+11, −8, −3	7	9
215	2	7 9— 7 48	753	+1, −3, +6, −4	4	12
216	2–3	11 52—12 21	766	+3, +5, −2, −7	4	12
217	3	3 13— 3 45	801	+13, −11, −1, −1	6	12
218	3	7 6— 7 35	776	+16, −23, +9, −2	12	12
219	6	7 44— 8 6	756	+6, −6, 0	4	9
220	7	7 8— 7 40	775	−20, −20, +23, +17	20	12
221	8	7 10— 7 43	785	+18, +1, −9, −11	9	12
222	9	7 15— 7 44	790	+23, −8, −5, −10	12	12
223	10	7 15— 7 56	763	+*20*, −33, −1, −4, +18	14	13
224	13	7 16— 7 43	771	−11, −10, +16, +5	10	12
225	14	7 7— 7 36	299,789	+28, −8, −16, −4	14	12

TABLE VI—continued

(1) Series	(2) Date	(3) P.S.T.	(4) Mean Velocity	(5) Residuals	(6) A.D. ±	(7) Wt.
	1933	h m h m	km/sec.			
226	Feb. 15	7 22— 7 50	299,767	+8, +*13*, −2, −19	10	11
227	16	7 10— 8 16	767	+18, +9, 0, −7, −5, −15	9	18
228	17	7 8— 7 52	779	+1, +*32*, +5, −35, −4	6	12
229	20	7 11— 7 47	782	+12, +7, −14, +5, −10	10	15
230	21	6 59— 7 28	784	+14, −21, +23, −16	18	12
231	22	7 22— 7 50	774	+36, −16, −14, −6	18	12
232	24	7 4— 7 46	807	−18, −15, +16, +17	16	12
233	27	7 5— 7 34	299,788	−4, +11, +9, −16	10	12

of the observations. Where inspection permitted or other notes supplied information, approximate times have been inserted. Hours between 12^h0^m and 4^h9^m are A.M.; all others are P.M.

Mean velocity (*col.* 4).—Each set of observations usually furnished three values of the velocity. Several sets were arbitrarily grouped into a series covering about an hour's time. Some special groupings will be noted, made either to cover a scattered series or to divide a night's readings into several small series. The velocity V given is the simple mean for the series.

Residuals (*col.* 5).—The residuals are the values which, applied to the mean velocity for a series, give the individual values for each set.

Average deviations (*col.* 6).—The mean without regard to sign of the residuals in column 5.

Weights (*col.* 7).—The weight is the number of single determinations of the velocity for the series. An italicized number in the residuals column indicates that the reading has a weight less than that of the normal value for the group. For series 1–25, 1931, when 10 *R* and 10 *L* readings were made, each set of 20 readings was given a weight of 1. In series 26–46, 1931, when 5 *R* and 5 *L* and then 5 *L* and 5 *R* readings were made, each set of 20 readings was given a weight of 2. From series 47, 1931, on, practically all the readings were made 5 *R*, 5 *L*, 5 *R*, 5 *L*, 5 *R*, and each set of 25 readings was given a weight of 3. In a few cases, the set of 25 readings being incomplete, weights of $\frac{1}{2}$, 1, and 2 were allotted, and in a few others, involving seven sets of 5 each, weight 5 was given.

The low rating given to the 1931 readings is due, first, to the way in which they were taken, which did not eliminate drift, and, second, to the fact that errors may have crept into the readings because the pendulum case was not controlled for temperature. The 1931 observations might have been considered as preliminary results and omitted altogether; but, owing to the large fluctuations in the individual values, it was decided to include every observation in the final mean velocity.

TABLE VII

Series	Date	No. Separate Determinations	Mean Velocity	A.D.
1– 54	1931 Feb. 19–July 14	493	299,770	±12
55–110	1932 Mar. 3–May 13	753·5	299,780	11
111–158	1932 May 13–Aug. 4	742	299,771	9
159–233	1932 Dec. 3—1933 Feb. 27	897	299,775	±11
		2885·5	299,774	±11

Table VII is a summary of the data given in Table VI. The average deviations in the last column are relative to the mean in the preceding column.

DISCUSSION

Distribution of velocities.—The mean velocities shown in column 4, Table VI, have been grouped into divisions covering a range of 5 km/sec and are shown in Table VIII. A plot of these data in Fig. 1.8 resembles a probability-curve and indicates that the probable value of a constant velocity would be 299,774 km/sec.

Time–velocity curves.—A plot of velocity readings with respect to time is shown in Fig. 1.9, the abscissae representing days of the year and the ordinates velocity. Four periods of the night are distinguished by the characters shown in the legend. The heavy line joins the weighted mean values of small groups of readings covering a few days' time, while the light line shows the axis drawn at a mean value of 299,774 km/sec.

All the 1931 observations lie close to the axis with the exception of series 14–25, whose mean is 299,746 km/sec. The 1932 curve begins

TABLE VIII
Frequency Distribution of Measured Velocities

Velocity Range	Number	Velocity Range	Number
299000+		299000+	
726–730	4	776–780	515
731–735	6·5	781–785	270
736–740	3·0	786–790	236
741–745	55	791–795	90
746–750	29	796–800	62
751–755	86	801–805	33
756–760	184	806–810	30
761–765	304	811–815	32·5
766–770	353·5	816–820	0
771–775	580	821–825	12

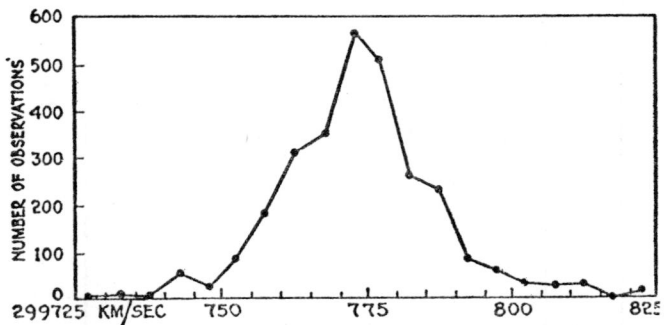

FIG. 1.8. Velocity–distribution curve.

at 299,800 km/sec, suddenly drops to 299,776 km/sec, continues just above the axis until early in May, then drops to 299,760 km/sec early in June. Several fluctuations occur in the curve at this time. The curve remains below the axis until the end of the observations on

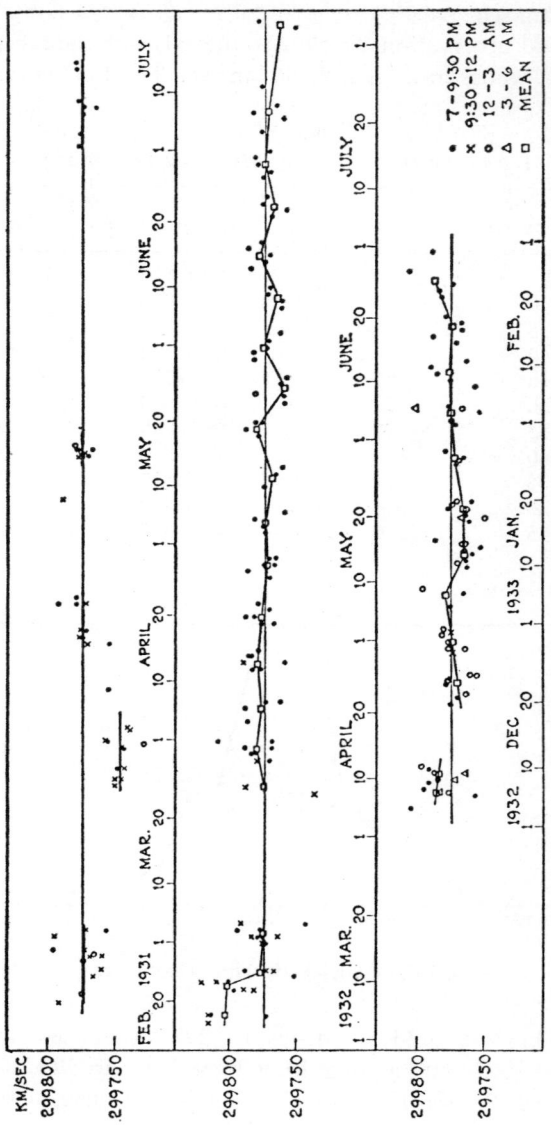

Fig. 1.9. Velocity–time curve.

August 4. The 1932–1933 curve begins at about 299,785 km/sec early in December, crosses the axis twice, and reaches a value of 299,765 km/sec on January 15. It then gradually rises to a value of 299,787 km/sec by the end of February.

Tidal-force–velocity curves.—When the time–velocity curves were first plotted, a curve freely drawn through the individual points appeared to resemble the tidal curve of the water depth at the nearby coast, if the tidal curve were displaced 10 hours forward. Since the

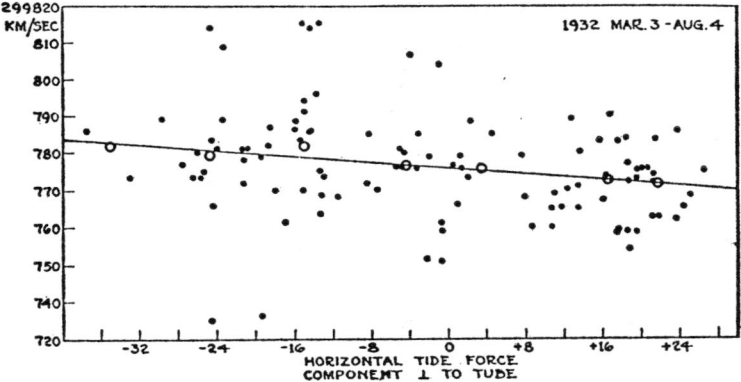

FIG. 1.10. Velocity–tidal-force curve.

lunitidal interval in this locality is 10 hours, the agreement seemed to suggest a relationship with sun–moon tidal forces acting at the time of observation and not with earth displacements due to the changing weight of water on the coast. The direct action of the tidal forces in producing earth expansion or in changing the period of the pendulum is too small to produce displacements of the order noted. Nevertheless, component curves of the sun–moon tidal forces were kindly drawn by the United States Coast and Geodetic Survey on their tide-predicting machine, from which values of component tidal forces were obtained for the times of observation. A slight correlation with the velocity is suggested in the case of the horizontal component perpendicular to the tube. The plot for early 1932 readings (Fig. 1.10) shows large velocities for a strong tidal force pulling in a southeasterly direction, and small velocities for a force directed northwesterly. A change of 10 km/sec corresponds to a change of $1 \cdot 35 \times 10^{-7}$ g. The scattering of the points is so large, however, that it is questionable whether the plot has any real significance.

Moon-diameter–velocity curves.—The diameter of the moon, which can be taken as a measure of its distance at the times of observation, was plotted against velocity for both the early 1932 and the 1932–1933 data (Fig. 1.11). Both curves show the same feature, namely, a curve

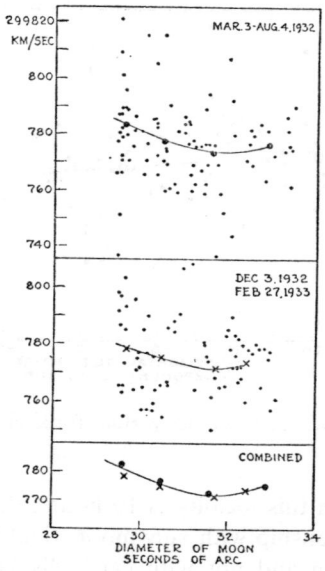

FIG. 1.11. Velocity–moon-diameter curve.

convex downward, almost coincident in the plots, indicating high velocities for both large and small diameters of the moon and suggesting tidal effects. The scattering, however, is large and the results of low weight.

Repeated measures of the base line and checks on the clock rate revealed nothing capable of accounting for the residual differences between the mean curve and the axis. A vibration of the mirror

2

The Velocity of Propagation of Electromagnetic Waves derived from the Resonant Frequencies of a Cylindrical Cavity Resonator*

BY L. ESSEN, D.SC., PH.D., AND A. C. GORDON-SMITH
The National Physical Laboratory

(*Communicated by Sir Charles Darwin, F.R.S.—Received 4 December,* 1947)

The frequency of resonance of an evacuated cavity resonator in the form of a right circular cylinder is given by the formula

$$f = v_0 \sqrt{\left[\left(\frac{r}{\pi D}\right)^2 + \left(\frac{n}{2L}\right)^2\right]\left[1 - \frac{1}{2Q}\right]},$$

in which v_0 is the free-space velocity of electromagnetic waves, D and L are the internal diameter and length respectively of the cylinder, r is a constant for a particular mode of resonance, n is the number of half-wave-lengths in the resonator and Q is the quality factor. Assuming the validity of this equation the value of v_0 can be obtained from measured values of f, D, L and Q.

A copper cylinder of diameter approximately 7·4 cm. and length 8·5 cm. was constructed with the greatest uniformity of diameter and squareness of end-faces and its dimensions were

---

* *Proceedings of the Royal Society*, volume A, 194, 1948.

system with a period equal to a fraction of that of the rotating mirror conceivably may have produced the rapid fluctuations observed in the individual readings. Further experiments on a more stable terrain, with improved self-recording apparatus, carried on continuously over an extended period of time will be necessary to clear up the problem.

measured. The resonant frequencies for a number of different modes were measured and experiments were made to show that the effects on frequency of the coupling probes to the oscillator and detector were negligibly small. It was concluded from these measurements that the most favourable experimental conditions can be obtained for the E_{010} and E_{011} modes. Final measurements on these gave

$$v_0 = 299{,}792 \text{ km./sec.}$$

The estimated maximum error of the result is 9 km./sec. (3 parts in 10^5). This is the error of a single measurement and, since most of the errors are not necessarily random, little is gained by making a large number of measurements. The value is 16 km./sec. greater than the recently determined values of the velocity of light, although the results are not in disagreement when the combined limits of accuracy are taken into account.

1. INTRODUCTION

The velocity of propagation of electromagnetic waves is, in Maxwell's theory, given by $1/\sqrt{(\mu\varepsilon)}$, μ being the permeability and ε the permittivity of the medium in rationalized practical units. The value for a vacuum, $1/\sqrt{(\mu_0\varepsilon_0)}$, is a constant of fundamental importance, being theoretically identical with the velocity of light, and appearing in a great number of electrical calculations. During recent years its value has also become of significance in the field of radio engineering, since it is required for the calculation of the resonant frequencies of electrical circuits, such as cavity resonators, and also for methods of navigation depending on radio waves. The measurement of this constant has attracted many research workers, and possibly more effort has been devoted to it than to the determination of any other general constant. In most of the experiments that have been performed the velocity of light has been determined directly by the measurement of the time occupied in its travel over a certain distance. In all cases the time interval is produced by a revolving or alternating mechanism so that the quantities actually measured are frequency and distance. Both of these can be measured with a high accuracy—better than 1 part in 10^5—and the main experimental difficulty appears to be the adjustment of the variable frequency or length to correspond to the fixed length or frequency as the case may be. In order to eliminate as far as possible the random observational errors a great number of determinations are necessary. Even so it was found in the two most recent experiments (Michelson, Pease & Pearson, 1935; Anderson, 1941) that the means of groups of a considerable number of measurements differed by amounts greater than would be expected if the errors were entirely random. Michelson *et al.* found differences of 11 km./sec. between the means of groups consisting of about 500 measurements and a maximum difference of 93 km./sec. between the means of groups of about six measurements. Anderson does not give the individual results, but he obtained a difference of 67 km./sec. between the means of groups of about 200 measurements. The close agreement between their final results of 299,774 ± 11 and 299,776 ± 14 km./sec. may therefore be misleading.

The constant has also been determined by a comparison between the calculated and measured values of the capacitance of a capacitor of suitable shape (Rosa & Dorsey, 1907). The value calculated from electromagnetic theory involves the permittivity of the medium and the dimensions of the capacitor. The measured value, obtained by balancing a bridge network of which the capacitor forms one arm, is expressed in terms of the international ohm. The experiment consists in the measurement of the dimensions of the capacitor, and of a resistance, a resistance ratio, and a frequency, and the result can be interpreted as the measured value of the permittivity of free space. From this the velocity can be calculated, the value of permeability being known by definition. Both the calculations and experiments are beset with difficulties, but the result obtained by Rosa & Dorsey was probably the most reliable up to that time. The final result given is 299,710 km./sec., which is the mean of about 900 individual determinations, the average deviation is 22 km./sec. and the estimated maximum error is 30 km./sec., apart from uncertainties in the value taken for the international ohm. This latter constant is now known with an accuracy of about 2 parts in 10^5, and Birge (1941) has applied a correction to their results to obtain a value of 299,784 km./sec. Using all the reliable data existing at the time Birge gives the most probable value for the constant as

$$c = 299,776 \pm 4 \text{ km./sec.}$$

In recent years a number of experimenters have measured the

speed of travel of electric waves of frequencies of the order of a few Mcyc./sec., but the accuracy achieved was in general inferior to 1% (Smith-Rose, 1943), so that these measurements cannot yet be taken into account in discussions on the value of the constant.*

The experiment described in this paper presents fewer technical difficulties than any method previously described. It consists in the measurement of the frequency of electrical resonance of a hollow copper cylinder, and the calculation of the frequency from electromagnetic theory. It is similar in principle, therefore, to the experiment of Rosa & Dorsey but also has some points in common with the velocity of light experiments in that the only quantities measured are the dimensions of the apparatus and a frequency. The frequency measured is, however, that of the electromagnetic wave itself and not that of a modulation imposed on the wave. Moreover, the frequency of the oscillations is of the order of 3×10^9 cyc./sec. compared with 5×10^{14} cyc./sec. for light vibrations. The dimensions and the frequency can be alined by setting to resonance with a precision of a few parts in 10^6 so that the accuracy of observation is of the same order as that of the measurement of the dimensions and frequency. Little is gained, therefore, by making a large number of determinations, and the authors feel justified in claiming a rather higher accuracy for a single measurement than is claimed by previous workers for their averaged results. The result obtained is 299,792 km./sec. This value is 1 km./sec. lower than that which has already appeared in Essen (1947), but the difference arises only from the fact that as the work proceeded the calculations were made to more significant figures than appeared to be warranted at first.

The experiment arose out of work on cavity resonator wavemeters, and some deductions concerning the velocity of propagation were made in a previous paper (Essen, 1946). The work has been done concurrently with an already full programme, and except for the cavity resonator itself existing equipment has been used. As the experiment progressed it became clear that further work with different modes of resonance and with cavities of different types would be of interest, particularly as the final result obtained is

---

* The accuracy has been greatly increased in works published since this was written and the probable error is now of the order of 1 part in 10^4 (Jones, 1947; Smith, Franklin & Whiting, 1947).

16 km./sec. higher than the previously accepted value of 299,776 km./sec. Such work would provide further experimental evidence concerning the extent of the different sources of possible error.

Plans have been made to continue the work as the opportunity arises, but in the meantime it is thought that the measurements already made are of sufficient interest to merit publication.

2. THEORY OF THE METHOD

The resonant frequency of a right circular hollow cylinder closed at both ends is given by Sarbacher & Edson (1943) as

$$f_{lmn} = v \sqrt{\left[\left(\frac{r}{\pi D}\right)^2 + \left(\frac{n}{2L}\right)^2\right]}, \qquad (1)$$

in which f_{lmn} is the frequency in cyc./sec. of the mode designated by the suffix, v is the velocity of propagation in cm./sec. in the medium ($v = 1/\sqrt{(\mu\varepsilon)}$), D is the diameter and L the length of the cylinder in cm., n is a whole number having the same value as the third suffix in the mode designation and r is the mth root of the Bessel equations $J_l(x) = 0$ for E modes and of $J'_l(x) = 0$ for H modes. If the cavity is evacuated $\mu = \mu_0$ and $\varepsilon = \varepsilon_0$ and the velocity is that for free space. We have then

$$v_0 = \frac{f_{lmn}}{\sqrt{\left[\left(\frac{r}{\pi D}\right)^2 + \left(\frac{n}{2L}\right)^2\right]}}. \qquad (2)$$

The modes of resonance which were finally used in the experiment were the E_{010} and E_{011} and for both of these modes the value of r is 2·404825, n being 0 and 1 respectively.

The above formula is derived on the assumption that the walls of the cavity are perfectly conducting. In practice, owing to the finite conductivity of the walls, the magnetic field penetrates them to some depth, and the field configuration therefore differs from that assumed in the case of a perfect resonator. No complete solution has yet been obtained for this case, but analogy with electrical circuits resonating at low frequencies suggests that the effective inductance of the system is increased as a result of the field inside the walls and that the resonant frequency is therefore reduced. Approximate expressions have been derived for the effect on the resonant frequency. Bernier (1946), for example, has shown that if the propagation constant in the lossless guide is β ($= 2\pi/\lambda = 2\pi f/v$, where λ is the wave-length in the guide), then the propagation constant in a guide of finite conductivity is

$$\beta' = \beta\left[1 - \frac{\mu eA}{2\sqrt{2}}(1-j)\right], \quad (3)$$

where A is a function of the field and e is the skin depth which depends on the conductivity in accordance with the well-known skin-effect formulae. The imaginary part of this expression corresponds to the attenuation, but the real part corresponds to a reduction in frequency. Putting the expression in terms of frequency we have

$$f' = f\left[1 - \frac{\mu eA}{2\sqrt{2}}(1-j)\right], \quad (4)$$

and the frequency measured will be the real part of this expression, $f\left(1 - \frac{\mu eA}{2\sqrt{2}}\right)$. This can be expressed in terms of the Q of the resonator which can be measured experimentally, because

$$Q = \frac{\text{real } \beta'}{2 \text{ imag. } \beta'} = \frac{1 - (\mu eA/2\sqrt{2})}{\mu eA/\sqrt{2}}, \quad (5)$$

and therefore $\quad f\left(1 - \frac{\mu eA}{2\sqrt{2}}\right) = f\Big/\left(1 + \frac{1}{2Q}\right), \quad (6)$

($\mu eA/2\sqrt{2}$) being a very small quantity, higher powers of which have been neglected.

For v_0 we must therefore use in place of (2)

$$v_0 = \frac{f'_{lmn}(1 + 1/2Q)}{\sqrt{[(r/\pi D)^2 + (n/2L)^2]}}, \quad (7)$$

f'_{lmn} being the measured frequency and Q the measured quality factor.

The measured Q's were rather lower than the calculated values, which is a general experience with cavity resonators. The reduction is probably due to either mechanical imperfections of the surface or an increased resistivity of the skin due to the mechanical treatment it receives. In any case it seems likely that the effect causing higher resistivity must also give a greater frequency correction and that the measured value should be used.

The resonator is coupled to the source of oscillations and to a detector by small probes through holes in one of the end-walls of the cylinder, and it is important to consider the effects of this coupling on frequency. The theoretical treatment would present considerable difficulties, but previous experimental results had indicated that the degree of coupling could be reduced until the effect on frequency was negligible. It was decided, therefore, to treat this as an experimental problem and to establish that the effect was negligible within prescribed limits of accuracy.

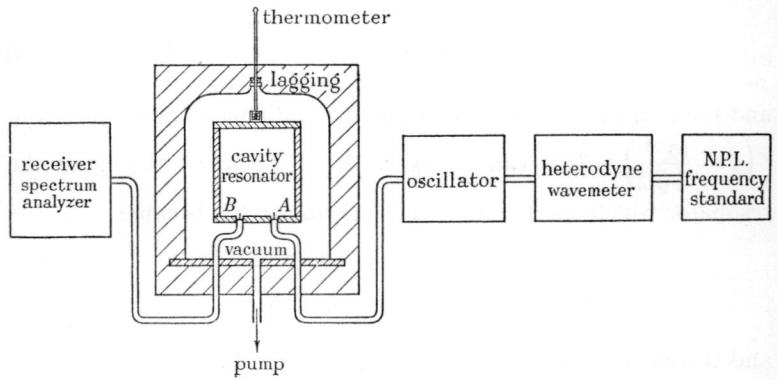

FIG. 2.1. Schematic representation of the experiment.

3. DESCRIPTION OF THE APPARATUS

3.1. *General arrangement*

The apparatus is shown diagrammatically in Fig. 2.1. The cavity resonator is coupled to an oscillator and a receiver by means of the probes A and B connected to coaxial lines. The frequency of the oscillator is varied, and when it reaches a value corresponding to one of the resonant frequencies of the cavity there is a sharp increase in the amplitude of the signal, indicated on a meter in the receiver. The frequency is set to give a maximum indication and its value is then measured by means of a heterodyne wave-meter. The resonator was enclosed under a bell-jar which could be evacuated, a free air path being left in the plug and socket fittings at A and B.

3.2. *The cavity resonator*

A number of factors were considered in the choice of the material, shape and size of the resonator. In order to obtain a sufficiently sharp resonance the inner surface, to a depth equal to the penetration of the field, must be highly conducting, but unfortunately highly conducting metals such as copper and silver are also soft and have a high temperature coefficient of expansion. They cannot therefore be worked with the highest precision, and the measurement of the dimensions of a resonator made from such materials presents some difficulty. The alternatives of plating a hard metal or fused quartz were considered, but it was thought that the plating could not be effected with the necessary precision and that the soft metal surface would still require to be worked. Most of the advantages would therefore be annulled, and the remaining one of low-temperature coefficient was not believed to be of paramount importance as temperature was unlikely to be a limiting factor in the accuracy of measurement. It was decided, therefore, to use solid copper as the material at least for the first experiments.

The frequencies of resonance can be calculated readily by existing theoretical methods for only simple shapes of resonator, such as the right circular hollow cylinder, the sphere and the rectangular parallelepiped.

Of these the last has some advantages, since it can be built from optically flat plates and its dimensions measured by interferometer methods, but it would be more difficult to make than the cylinder, and the large number of sharp edges and joints is a disadvantage. The sphere has the advantage of freedom from sharp edges but is not easy to make with the required precision. It was decided, therefore, to use the cylindrical form for the first experiments. There is considerable latitude in the choice of the dimensions of the resonator and the resonant frequency. Oscillators are available for any frequency up to 10,000 Mcyc./sec. and the skin-effect correction decreases with increase of frequency. On the other hand, the power and frequency stability of the oscillators both decrease with increase of frequency, particularly at values much higher than 3000 Mcyc./sec. The optimum size of the cylinder from the point of view of precision construction, with existing honing equipment, was of the order of 8 cm. both in diameter and length. Such a cylinder gives low-order resonances in the region of 3000 Mcyc./sec., and it appeared therefore that this was the best frequency to use. The exact diameter and

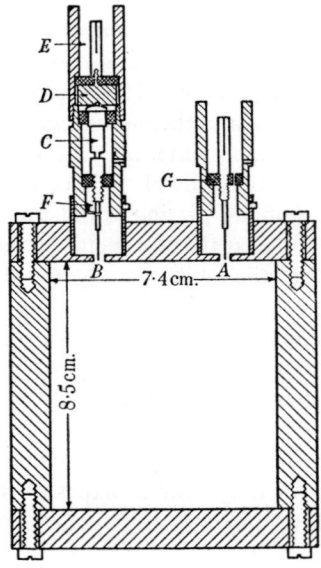

FIG. 2.2. Details of the resonator.

length were then chosen so that several low-order modes of resonance could be obtained in this region of frequency but spaced sufficiently to be regarded as single resonances in the theoretical treatment.

Details of the resonator are shown in Fig. 2.2. The cylinder is turned from the solid and then honed to be as uniform in dimensions as possible, the exact size not being important. The end-faces are ground square and parallel. The end-plates are ground flat and fixed to the cylinder by eight screws, an ample clearance being allowed in the holes in the plates. The coupling probes A, B pass through holes in one of the end-plates, the depth of insertion being adjustable. In the final condition no. 38 s.w.g. wire (diameter 0·0060 in. $\equiv$ 0·0152 cm.) was used through holes of 0·03 cm. diameter. Probe A is connected to a socket in which is plugged the coaxial lead to the oscillator. If the superheterodyne receiver is used probe B is terminated in a similar socket, but if the detecting system is a crystal detector and galvanometer the crystal is connected between the probe and the socket as shown at C in the diagram. The metal piece D acts as a radio-frequency by-pass condenser and the galvanometer is plugged into the socket E. The circuit requires a direct current return lead which is provided by a wire F, the position of which can be altered to tune the crystal circuit. The insulating material G and the sockets are drilled to permit the free passage of air during the evacuation of the system.

3.3. *The receiver*

The crystal detector used in the early experiments was a standard cartridge mounted silicon crystal with a tungsten wire (type no. CV 113), and its direct current output was observed on a Tinsley type 4500 A (50 ohms) galvanometer. For the final measurements it was decided to use instead a superheterodyne receiver which was available and which proved to be slightly more sensitive. This is a special form of instrument (known as a spectrum analyzer or spectrometer) developed for observing the spectral distribution of energy in the output of transmitting valves. Its useful feature in the present application is that its sensitivity remains almost constant, without adjustment of any controls, over a small band of frequencies. The tuning controls of the receiver are set so that the resonant frequency of the cavity is near the middle of this band. Then as the oscillator frequency is varied through the resonant value a peak deflexion is observed without any retuning of the receiver.

3.4. *The heterodyne wave-meter*

This instrument and its method of use have already been described by the authors (1945). With careful operation it enables frequencies in the region concerned to be measured with an accuracy of ± 2 parts in 10^6.

3.5. *The oscillator*

The oscillator valves were chosen from those used as local oscillators in receivers, the type CV 35 being suitable for the E_{010} mode and the CV 234 for the E_{011} mode. The former was operated from a stabilized mains supply unit and the latter from batteries. The tuning mechanisms supplied with the valves were found to give a sufficiently smooth frequency variation, although it was sometimes necessary to choose a good portion of the range of the tuning mechanism by the appropriate adjustment of subsidiary tuners.

4. MEASUREMENT OF RESONANT FREQUENCIES AND QUALITY FACTORS

The assembly of resonator and bell-jar was well lagged to ensure that any temperature changes occurring were very slow, and that the value recorded on a thermometer in good thermal contact with the resonator represented its actual temperature within $0.1°$ C. The system was evacuated at least 12 hr. before the electrical measurements were made. The electrical equipment was also switched on some hours before it was required for use in order that the various oscillators would reach a steady temperature and the measurements could be made under the most favourable conditions.

As explained in § 3.1 the measurement consisted in the adjustment of the oscillator to the resonant frequency of the cavity and then in the measurement of this frequency in terms of the standard. It was carried out by two observers to eliminate any appreciable time interval between the two operations. Precautions were taken to ensure that the maximum deflexion in the receiver did in fact correspond to the resonant condition of the cavity.

It was established that there was no appreciable change in the output of the oscillator as the tuning was varied in the region of resonance. There are in the circuit several short lengths of cable and connecting plugs and sockets. The impedance of these and therefore the power available at the probes changes with frequency, but owing to the resistance of the cable such tuning effects were of a different order of sharpness from the cavity resonance. It was easy to arrange that for the small frequency change involved in passing through resonance there was no appreciable amplitude change due to these effects. An overall check was obtained, in preliminary experiments, by increasing the insertion of the probes in order to give a convenient deflexion on the oscilloscope of the receiver by direct coupling between them when the cavity was not in resonance. The frequency was varied a few megacycles on both sides of the resonant value, and it was verified that the amplitude of the oscilloscope deflexion remains constant as it moves along the time base on both sides of the cavity resonance.

The voltage supplies to the oscillator were set so that frequency variations due to any residual fluctuations in them were reduced to a minimum. Under good conditions the frequency band over which an audible beat note could be obtained in the heterodyne wave-meter was less than 1 part in 10^5 of the frequency, and the setting to the mid-point of the band was made with a precision of $\pm$ 2 parts in 10^6.

The Q of the resonator was measured by observing the change in frequency from the resonant value required to reduce the detected current to $1/\sqrt{2}$ of its peak value. If this change is δf then the Q is $f/2\delta f$. The value was checked by measurements made with the crystal detector and galvanometer under conditions for which the detector was known to give a square law within close limits.

5. MEASUREMENT OF DIMENSIONS

The dimensions of the resonator were measured in the Metrology Division of the National Physical Laboratory.

5.1. *Internal diameter*

The measurements were made in a temperature-controlled room with a horizontal comparator arranged for internal measurements through steel contact tips of rounded form (about 0·4 cm. radius) operating under a measuring force equivalent to 0·34 kg. weight. A suitably wrung combination of slip-gauges and end-pieces served as the reference basis for internal comparison. Differences between this and any particular diameter of the tube were indicated on the optical scale of the comparator which could be read directly to $1·3\mu$ and by estimation to $0·3\mu$ (0·00001 in.). The measurements were made over eight symmetrically disposed diameters at seven positions along the axis of the cylinder.

The accuracy of determination of the mean measured diameter was estimated at $\pm 1\cdot3\mu$ for the preliminary measurements which were carried out as normal routine tests. To improve this for the final measurement the following modifications were made in the technique of measurement:

(a) The comparator was screened by means of a glass panel so that temperature effects due to the presence of the observer were minimized as much as possible.

(b) An improved form of reference standard was used. This was of the 'box' type and consisted of a firmly wrung combination of two piles of high-quality slip-gauges not differing in length by more than about $0\cdot03\mu$, between two lapped end-pieces, thus constituting a rigid 'box', or frame for which the reference dimension was well established. Each slip-gauge employed in the combination had previously been calibrated by interferometer measurements in terms of wave-lengths of light to an accuracy equivalent to $0\cdot03\mu$.

(c) Two diameters at the centre of the copper cylinder and in directions at right angles to one another were accurately measured in terms of the reference standard.

(d) The temperatures of the copper cylinder and the hardened steel standard were observed and the results of the intercomparison were reduced to 20° C, using thermal coefficients of expansion of 12×10^{-6} and 17×10^{-6} for hardened steel and copper respectively.

(e) Having established two basic diameters all other diameters of the cylinder were compared with first one and then the other of the basic diameters. The effect of any temperature variations on the results obtained during each of these long series of comparisons was thereby eliminated provided any changes in temperature were reasonably the same for all parts of the cylinder. The mean measured diameter derived from each of the two series agreed to $0\cdot5\mu$.

(f) Theoretical and experimental investigations were made of the different effects of the compressive force exerted by the steel measuring tips of the comparator on hardened steel and copper surfaces. The results obtained by calculation and experiment agreed to $0\cdot03\mu$ and showed that the difference could be compensated by reducing the observed value of internal diameter of the cylinder as derived from the standard by $0\cdot27\mu$.

5.2. Length

Length measurements were made by means of a vertical comparator having a rounded contact measuring tip and an optical scale similar to that of the horizontal comparator. One end of the cylinder was placed in contact with the horizontal platen of the comparator under the contact tip, and the height of the other end of the cylinder was measured at various positions by comparison with a wrung combination of reference standard slip-gauges. Under these conditions of measurement it was considered that the compressive effect of the single contact tip was negligible in relation to the general accuracy of the length determination.

The overall length of the tube was thus measured at eight positions symmetrically spaced around the cylinder, and at each of these positions three measurements were made, one near the bore, one at the mid-point of the wall thickness and one near the outside of the tube. The measurements are considered to be reliable to $\pm 0.8\mu$.

Some tests were made to show whether the diameter and length were affected by screwing on the end-plates, but within the accuracy of these measurements ($\pm 0.8\mu$) there was no evidence of deformation.

6. EXPERIMENTAL RESULTS

6.1. Preliminary measurements

The method of coupling used favours the excitation of the E_{01n} modes but the H_{011} and H_{111} modes can be excited if the probes are bent round to form small loops orientated so as to give a linkage with the field associated with the particular mode. To gain experience of the most suitable experimental conditions preliminary measurements were therefore made of the resonant frequencies of a number of different modes.

From the frequencies the values for velocity were calculated and the results are included in Table 1, but they are not considered to be more accurate than 6 parts in 10^5 for the following reasons:

(a) The metrological examination of the cylinder revealed larger variations of diameter than anticipated, and the measurements themselves were not made with the maximum possible accuracy, the uncertainties being given at the head of Table 1. The end-faces were not ground flat and the length measurement is therefore averaged to allow for this, the deviations being of the order of $\pm 5\mu$.

(b) The conditions for the frequency measurement were not ideal. For the E_{012}, H_{011} and H_{111} modes the coupling probes were inserted about 1 or 2 mm., and in the case of the two latter modes were formed into small loops. The size of the coupling holes was 0·15 cm.

TABLE 1

(*Preliminary Measurements*) *Resonant Frequencies of the Copper Cylinder and Calculated Values of Velocity*

Mean diameter of resonator 7·39759 ± 0·00012 cm.
Variation ± 0·00045 cm. Length of resonator 8·55844 ± 0·0003 cm.

Mode of resonance	Correction factor $(1 + 1/2Q)$	Constant r	Frequency f_0' (Mcyc./sec.)	v_0 (km./sec.)
E_{010}	1·000028	2·404825	3102·12	299,797
E_{011}	1·000035	2·404825	3562·38	299,798
E_{012}	1·000030	2·404825	4678·81	299,785
H_{011}	1·000015	3·831706	5243·96	299,799
H_{111}	1·000031	1·841184	2950·78	299,777

(c) As a result of these measurements it was decided to concentrate on the E_{010} and E_{011} modes, which could be detected readily with the probes flush with the wall of the cavity. The size of coupling hole was reduced. It was at the same time decided to regrind the cylinder to improve its uniformity of diameter and length.

6.2. *Effect of size of coupling hole and the penetration of the probes*

It was necessary to establish experimentally the effect on frequency of the presence of the coupling holes and probes. First measurements were made with three sizes of coupling hole, the probes being in each case as nearly flush as possible to give an adequate indication. The results are given in Table 2. It was concluded from these results that any residual effect on frequency was less than 0·01 Mcyc./sec., i.e. less than 3 parts in 10^6.

The effect of probe insertion is shown in Figs. 2.3 and 2.4, and from these results it is concluded that the residual effect of the probes is again less than 0·01 Mcyc./sec.

TABLE 2

Mode	Size of coupling hole (cm.)	Insertion of probe (cm.) (estimated by eye)	Resonant frequency (Mcyc./sec.)
E_{010}	0·03	< 0·01	3101·246
	0·07	flush	3101·251
	0·16	flush	3101·242
E_{011}	0·03	< 0·01	3563·798
	0·07	flush	3563·787
	0·16	flush	3563·71

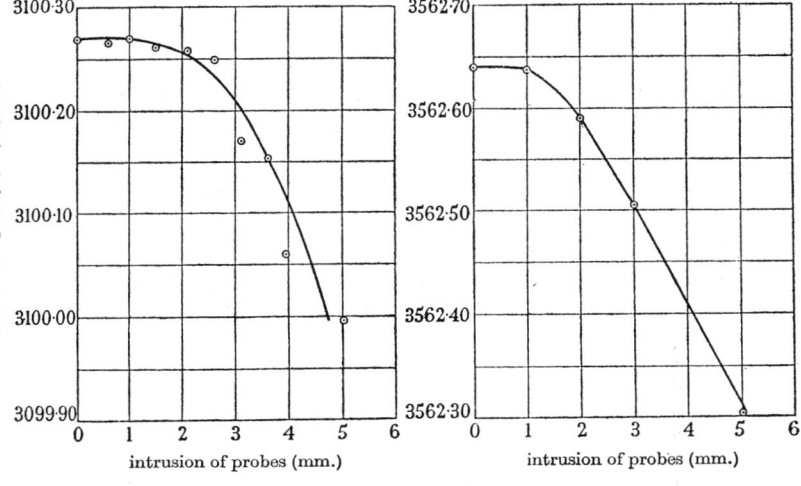

FIG. 2.3. E_{010} mode (in air).   FIG. 2.4. E_{011} mode (in air).

FIGS. 2.3 and 2.4. Effects of probe intrusion. Diam. of holes 0·03 cm. Diam. of probes 0·015 cm.

TABLE 3

Diameter of the Resonator Expressed as the Difference from the Mean Value of 7·39957 *cm. Unit* 0·00001 *cm.* (0·1μ)

Axial position of measurement from one end (cm.)	Measured diameter at 20° C in different diametrical planes							
	1	2	3	4	5	6	7	8
0·13	− 15	− 6	+ 1	+ 1	+ 7	+ 3	0	− 10
1·3	− 20	− 11	− 4	− 2	+ 1	+ 2	− 4	− 16
2·8	− 20	− 13	− 17	− 2	+ 9	+ 7	− 4	− 20
4·3	− 15	− 9	− 6	+ 3	+ 8	+ 10	− 3	− 14
5·8	− 2	+ 6	+ 13	+ 13	+ 27	+ 24	+ 13	0
7·3	− 3	− 3	+ 6	+ 10	+ 14	+ 9	+ 7	0
8·4	+ 7	− 1	+ 5	+ 8	+ 10	+ 8	+ 2	+ 9

TABLE 4

Length of the Resonator Expressed as the Difference from 8·53637 *cm.* (*Mean of Inner Edge Values*). *Unit* 0·00001 *cm.* (0·1μ)

Position of measurement	Mean overall length at 20° C		
	near inner edge	centre	near outer edge
1	− 8	− 13	− 16
2	+ 5	+ 2	− 8
3	+ 9	+ 6	+ 1
4	+ 9	+ 9	+ 3
5	+ 7	+ 2	− 3
6	− 2	− 5	− 12
7	− 13	− 14	− 17
8	− 9	− 13	− 17

The Q of the resonator for the two modes had already reached its maximum value with the coupling considerably closer than that finally used. The values are 18,000 and 14,000 respectively for the E_{010} and E_{011} modes as compared with the theoretical values of 21,000 and 17,600. The ratios of experimental to theoretical values are as great as those usually obtained for cavity resonators.

6.3. *Final measurements*

The final results for the velocity are based on a series of metrological measurements made between two series of electrical measurements. The detailed dimensions are given in Tables 3 and 4. The cylinder is seen to be slightly oval in shape, the difference between the maximum and minimum diameters being 2μ (3 parts in 10^5). Since the field for the modes employed is symmetrical about the axis, the effective diameter was taken as the mean value. The length at the inner edge of the cylinder was slightly greater than that at the outer edge and varied smoothly round the circumference by a total of $2 \cdot 2\mu$. The effective length was therefore taken as the mean value at the inner edge.

The measured values of resonant frequency and the values of velocity calculated from the resonant frequencies and dimensions are given in Table 5. The frequencies were measured using the minimum coupling conditions as follows:

Size of coupling holes	0·03 cm.
Diameter of probes	0·015 cm.
Insertion of probes	< 0·01 cm.

Resonance was detected on the spectrum analyzer as described in § 3.3.

The estimated maximum errors due to various causes are listed below:

(1) Setting of the frequency to resonance and measurement of the frequency	$0 \cdot 4 \times 10^{-5}$
(2) Uncertainty of temperature of the resonator	$0 \cdot 2 \times 10^{-5}$
(3) Dimensional measurements	$0 \cdot 3 \times 10^{-5}$
(4) Residual effects of coupling holes and probes	$0 \cdot 6 \times 10^{-5}$
(5) Non-uniformity of the resonator	$1 \cdot 0 \times 10^{-5}$
(6) Uncertainty of Q	$0 \cdot 5 \times 10^{-5}$
Estimated maximum error	3×10^{-5}

The uncertainties due to the first three causes can be established from the accuracy of repetition and a knowledge of the band-width of the receiver and frequency stability of the oscillator. The errors due to the other causes cannot be estimated with such certainty. That due to the coupling is estimated from the results given in § 5.2. As regards the non-uniformity of the resonator it will be seen from Tables 3 and 4 that the maximum deviations of the dimensions from

TABLE 5

Final Values for Resonant Frequencies of the E_{010} and E_{011} Modes and the Deduced Values of the Velocity of Propagation

Length of resonator 8.53637 ± 0.00008 cm.
Diameter of resonator 7.39957 ± 0.00003 cm.

Date	Mode of resonance	Correction factor $(1 + 1/2Q)$	Constant r	Measured frequency f'_{lmn} (Mcyc./sec.)	v_0 (km./sec.)
2. x. 46	E_{010}	1·000028	2·404825	3101·25	299,793
2. x. 46	E_{011}	1·000035	2·404825	3563·80	299,791
25. x. 46	E_{010}	1·000028	2·404825	3101·28	299,796
25. x. 46	E_{011}	1·000035	2·404825	3563·77	299,789

Mean value of velocity 299,792 km./sec.

the mean are $+ 2.7$ and $- 2.0\mu$, and that the variations occur in a smooth manner. It is estimated therefore that the average dimensions are known to 0.5μ. Previous results (Essen, 1946) obtained with much less uniform resonators support the assumption that an average value can be taken, although it cannot be assumed that the non-uniformity of dimensions affects different modes of resonance in the same way. An overall error of 1×10^{-5} has therefore been allowed for these uncertainties.

The total value of the correction due to the finite resistance of the walls of the resonator is approximately 3×10^{-5}. An uncertainty is introduced because of the imperfections in the surface of the resonator and a lack of knowledge of the precise path of the current. It is thought that these effects are to some extent taken account of by using the measured value of Q in the calculations. The value of Q is accurate to about 5%, and the overall error has been estimated at 0.5×10^{-5}.

The results obtained before and after the metrological measurements do not give quite such good agreement as was usually obtained during the preliminary measurements on dismantling and reassembling the resonator. However, in view of the importance of associating the frequency measurements closely in time with the measurement of dimensions, the values were accepted and their mean taken as giving the final result. The difference between the values of velocity obtained for the two modes is also rather larger than expected. The uncertainty in the dimensions could contribute to this, since the length and diameter affect the resonant frequencies to different extents, and the residual errors due to coupling and non-uniformity could also have different effects on the two modes.

The work described above was started in the Radio Division of the National Physical Laboratory as part of the programme of the Radio Research Board and completed in the Electricity Division as part of the research programme of the National Physical Laboratory. This paper is published by permission of the Department of Scientific and Industrial Research.

The authors wish to acknowledge the assistance of Mr H. Barrell of the Metrology Division who was responsible for the measurement of the dimensions of the resonator, and of Mr J. W. Simmons of that Division who was responsible for its construction.

REFERENCES

ANDERSON, W. C. (1941), *J. Opt. Soc. Amer.* **31**, 187.
BERNIER, J. (1946), *Onde Elect.* **26**, 305.
BIRGE, R. T. (1941), *Physics*, **8**, 90.
ESSEN, L. (1947), *Nature*, **159**, 611.
ESSEN, L. (1946), *J. Instn Elect. Engrs*, **93**, part III A, no. 9, p. 1413.
ESSEN, L. and GORDON-SMITH, A. C. (1945), *J. Instn Elect. Engrs*, **92**, 291.
JONES, F. E. (1947), *J. Instn Elect. Engrs*, **94**, part III, p. 399.
MICHELSON, A. A., PEASE, F. G. and PEARSON, F. (1935), *Astrophys. J.* **82**, 26.
ROSA, E. B. and DORSEY, N. E. (1907), *Bull. U.S. Bur. Stand.* **3**, 433.

SARBACHER, R. I. and EDSON, W. A. (1943), *Hyper- and Ultra-high Frequency Engineering*. London: Chapman and Hall.

SMITH-ROSE, R. L. (1943), *J. Instn Elect. Engrs*, **90**, part I, no. 25, p. 29.

SMITH, R. H., FRANKLIN, E. and WHITING, F. B. (1947), *J. Instn Elect. Engrs*, **94**, part III, p. 391.

3
A Determination of the Velocity of Light*

By Erik Bergstrand

INTRODUCTION

Fizeau's method of determining the velocity of light is based on the fact that a light signal takes a certain time to cover a known distance. For geodetical purposes an exact knowledge of the velocity of light is of immediate interest owing to the circumstance that, the velocity being known, the inversion of Fizeau's principle will determine a distance. Here, instead of light, radar pulses may be considered just as well. In a vacuum there is no reason to assume a difference in velocity between light and radar pulses. For an accurate determination light signals, which have but little scattering and bending, are to be preferred.

In 1941 with the aid of a grant from the Längman Foundation the writer started some preliminary experiments according to Fizeau's principle at the Nobel Institute of Physics. The experiments were continued, still at the Institute, with the aid of grants from the Swedish State Council of Technical Research, and in 1947 they resulted in an apparatus for "Measurement of distances by high frequency light signalling." (18)† Tests were carried out at Lovö, close to Stockholm. The results were so promising that The Geographical Survey and the Gasaccumulator Co, Ltd (AGA), together paid for a new technically improved and more sturdy apparatus. It was built by AGA and included a lot of the actual parts from the old one. In the autumn of 1948 a "Preliminary determination of the velocity of light" (14) was carried out by the not yet fully completed AGA-model, the "geodimeter." During the winter of 1948–49 the

---

* *Arkiv för Fysik*, volume 2, 1950. † Ref. p. 142.

geodimeter was ready. The velocity measurements described below were performed in the spring and autumn 1949 at the 7 km. Geographical Survey base line near Enköping. In August measurements of distances were effected in Norrland.

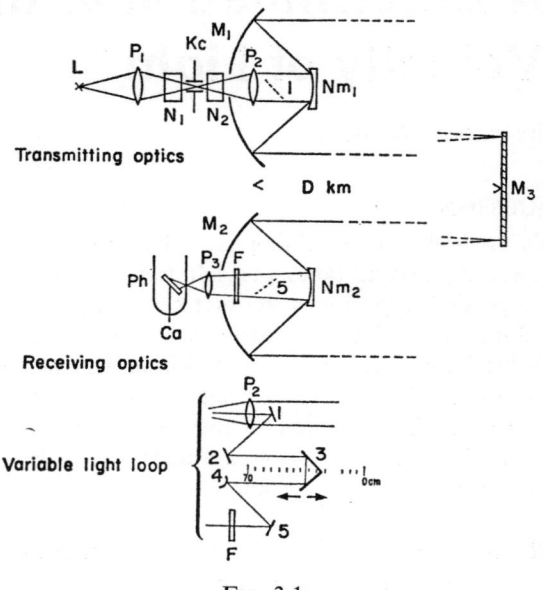

Fig. 3.1.

GENERAL DESCRIPTION OF THE APPARATUS AND ITS EMPLOYMENT

Figures 3.1 and 3.2 show the chief parts of the geodimeter and their use.

By means of a lens, P_1, the image of a small constant light-source, L, is projected between the plates of a Kerr cell, Kc. On each side of the cell there are the crossed Nicol prisms, N_1 and N_2. In front of the prism N_2 the intensity of the light depends on the tension between the plates in accordance with the equation:

$$J = J_0 . \sin^2 k . V^2 \qquad (1)$$

where J = intensity, J_0 and k = constants, V = difference of potential between the plates. The intensity of the light as a function of the tension is shown by Fig. 3.3. Between *a* and *b* the curve is almost straight. If the cell has a constant bias tension of rest between these values and an alternating tension of limited amplitude is superimposed, the intensity curve of the light is almost identical with that of

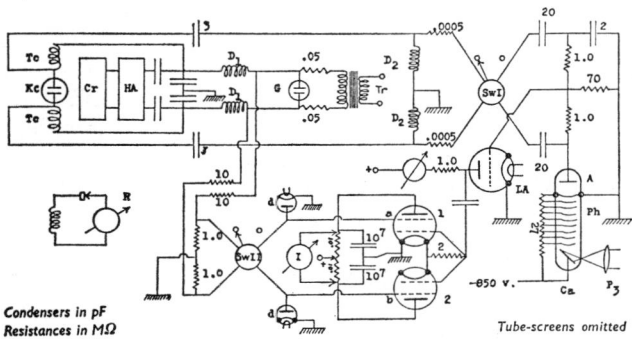

Fig. 3.2.

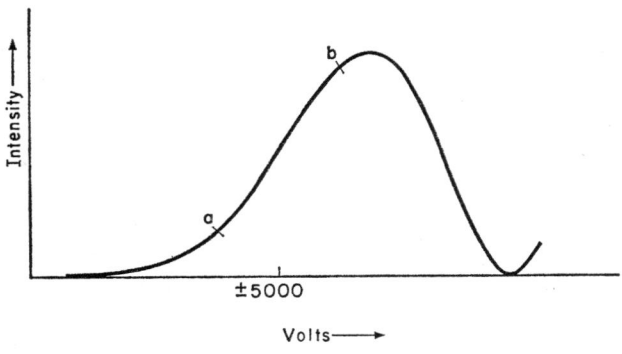

Fig. 3.3.

the alternating tension. Added to this is a quantity of constant light depending on the bias tension. Via the tuning coils, Tc (Fig. 3.2), and the amplifier, HA, the plates of Kc are fed with high-frequency tension from the crystal oscillator Cr. The amplitude is 2000 volts and the frequency 8·33 Mc. Instead of the constant bias tension there is a 50-cycle alternating tension with an amplitude of 5000 volts. The 50-cycle oscillation curve being rectangular in shape (⎍⎍⎍), it gives the most favourable conditions. The result is a completely constant tension of rest half-way between a and b, and this tension changes its sign 100 times a second. A reasonably good rectangular outline is obtained by means of a chain of neon tubes G and the resistances ·05 placed over the secondary coil of the transformer Tr. Thus the tops of the sine tension are cut off. The chokes D (throttles) prevent the high frequency from short-circuiting. From the size of the amplitudes and Fig. 3.3 it is evident that the high frequency variations change their phase simultaneously with the change of sign of the Kc-poles. This takes place 100 times a second owing to the feeding from Tr.

The transmitting optics in Fig. 3.1 project the image of L to the distant area where the plane mirror M_3 is set up. M_3 reflects the light straight back and thus a small part of the image of L falls upon the receiving optics, where it is absorbed by the phototube Ph. Ph is a multiplier cell, fed with high frequency by a branch from the line to Kc (Fig. 3.2). The small condensers 3 and the H.F.-chokes D_2 keep the 50-cycle tension away from the phototube. The resistances 1, the condenser 2 and the anode capacity of the cell constitute a bridge preventing high-frequency tension from reaching the amplifier LA. The phase of the high-frequency feeding can be changed by means of the pole-switch SwI.

Owing to the rate of the feeding tension being in phase with the incoming light variations, the photo current will give rise to a series of more or less strong rectified 50-cycle voltage pulses in the resistance 70. These pulses may have different magnitudes among themselves, all depending on whether the entering light has been emitted during the negative or positive half-cycles of the 50-cycle polechanges, for in the two series of half-cycles the light variations have, of course, opposite phases. The tension pulses are amplified by LA and then imposed on the control grids of the valves 1 and 2. Via the resistances 10 and then the diodes d the suppressor grids (a and b) of 1 and 2 are fed by 50-cycle tension from Tr in such a manner that a has a normal working tension at the same time as b has a strong negative tension and vice versa. Thus valve 1 only transmits the one series of tension pulses owing to light issued when the poles of Kc have a certain sign position. Valve 2 only transmits the second series of voltage pulses. As the oil-damped (10 sec.) direct current μA-meter, I, shows 0-deflection, the two series of pulses are of equal strength or, if all adjustments are not quite correct, a difference of current equal to that obtained when M_3 is screened off. This state settles the 0-deflection. Owing to the existing frequency of 8·332 Mc, I shows 0-deflection every increase of nine meters in the distance from the mirror M_3. Using equations we get the following outline:

At the plates of the Kerr cell the high-frequency tension is assumed to have the equation:

$$V = a \cdot \sin \omega t \tag{2}$$

a = amplitude, $\omega = 2\pi n$, where n = frequency, t = time.

It takes a time t_1 to transform V into light variations just outside the prism N_2. Considering a–b in Fig. 3.3 to be a straight line, we can express the light intensity by

$$J_{50+} = C_1 + C_2 \cdot \sin \omega (t - t_1) \tag{3 a}$$

$$J_{50-} = C_1 + C_2 \cdot \sin [\omega (t - t_1) + \pi] \tag{3 b}$$

50^+ and 50^- indicate that the light is emitted during the positive or negative half-cycles of the 50-cycle alternation. C_1 is the constant intensity and C_2 the amplitude of the variations. The angle π in (3 b) is due to the pole change.

When it has passed the distance D to the plane mirror M_3 and back again, the light or part of it is absorbed by the phototube. During the positive *high frequency* half-cycles we get current from the anode. Assuming early saturation, we get the photo-current:

$$i_{50+} = \tfrac{1}{2} \cdot n \cdot \int_{t_3}^{t_3 + \frac{1}{2n}} [A + B \cdot \sin \omega (t - t_1 - t_2 - 2D/c)] \cdot dt \tag{4 a}$$

$$i_{50-} = \tfrac{1}{2} \cdot n \cdot \int_{t_3}^{t_3 + \frac{1}{2n}} [A + B \cdot \sin (\omega (t - t_1 - t_2 - 2D/c) + \pi)] \cdot dt \tag{4 b}$$

A and B correspond to C_1 and C_2 of eq. (3). In the time quantity $2D/c$ (c = velocity of light), $2D$ is the whole distance from the prism N_2 (or, more exactly, from the point where t_1 in eq. (3) was reckoned) to the mirror M_3 and exactly the same distance back again. The cathode of Ph is assumed to be here. t_2 is practically the running time of the photoelectrons ($2 \cdot 4 \times 10^{-8}$ s.) from cathode to anode. t_2 may be considered to include a further small quantity of time, due to the cathode not being placed at the end of the distance $2D$. The summation in eq. (4) is done for one positive high frequency half-cycle, which begins at t_3 and ends at $t_3 + 1/2n$, t_3 being the transferring time of the tension from the Kerr plates to the photo anode and n the frequency. The charge per second or the current is obtained after multiplying by $n/2$, the number of half-cycles in one second of 50^+ and 50^- respectively.

The two currents j_{50+} and i_{50-} have opposite influences on the instrument **I**, which thus makes a deflection proportional to the difference of the currents. From eq. (4) we obtain:

$$i = i_{50+} - i_{50-} = \frac{B}{\pi} \cdot \cos \omega (t_3 - t_1 - t_2 - 2D/c) \qquad (5)$$

The current will be $i = 0$ to values on D, which fulfil the condition:

$$\omega \cdot (t_3 - t_1 - t_2 - 2D/c) = \frac{\pi}{2} - N \cdot \pi \qquad (6)$$

where N denotes a whole number. With $\omega = 2\pi n$ and $c = \lambda \cdot n$ where λ = the wave length, we get:

$$D = \frac{n}{2}(t_3 - t_1 - t_2) \cdot \lambda + \frac{2N-1}{8} \cdot \lambda * \qquad (7)$$

or

$$D = K + \frac{2N-1}{8} \cdot \lambda \qquad (8)$$

where K is a constant.

---

* Without making any restrictive assumptions about amplitudes or rectilinearness Mr. J. CLENDINNING (priv. com.) has proved the general validity of the eq. (7) by putting the total applied voltage in the place of V in eq. (1) and after expanding he thus obtains eq. (5) in mere terms of the form: $R \cdot \cos p \cdot \omega (t_3 - t_1 - t_2 - 2D/c)$. R = num. const., p = an odd integer.

D are distances to the plane mirror. In the case of a difference between two D, K will be eliminated and we get:

$$D_N = N \cdot \frac{\lambda}{4} \qquad (9)$$

D_N being known we get λ. N is easy to determine from approximate values. The frequency corresponding to λ being n and known, we get the velocity in air from:

$$c_a = n \cdot \lambda \qquad (10)$$

The actual frequency of 8·332 Mc makes $\lambda = 36·0$ m.

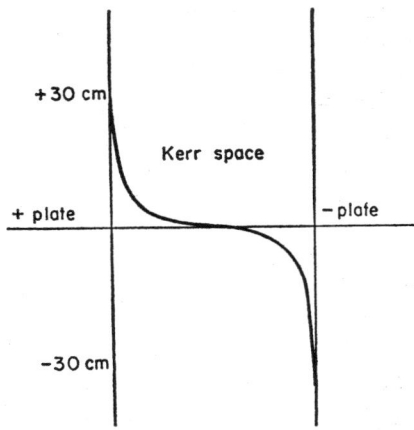

Fig. 3.4.

In practice the distant mirror is placed close to a zero-point (as read off on the μA-meter). The constant crystal frequency is changed by a small ($< 0·1°/_{00}$) definable amount until zero-deflection is reached.

This single reading off is impaired by a systematic unknown error, dependent on the location of the mirror M_3 in the image of the light source. Different parts of the image have different intensities, dependent on the part from which the light is emanating in the space between the Kerr plates. In Fig. 3.4 is shown the manner in which the change in distance corresponding to the variation of intensity varies from the positive to the negative Kerr plate.

The curve is obtained at an appropriate distance D by progressively placing a very small mirror in different parts of the image of light.

The error will not be eliminated by the change of poles, because of the simultaneously changing of phase of the valves 1 and 2 (Fig. 3.2). However, we get two distances D if the wires to the Kerr plates are shifted, and by taking the mean we get a correct value because of the inverse symmetry of the curve in Fig. 3.4. In this way the error will be eliminated. Instead of a shifter at the high-tension poles of the Kerr cell, two low-tension shifters SwI and SwII are inserted, one in the high-frequency feeding circuit to the photocell and the second in the 50-cycle circuit to the grids a and b (Fig. 3.2). Thereby a distance D is determined from the mean of four determinations corresponding to the four possible combined positions of the shifters SwI and SwII.

To obtain the difference of two D's according to eq. (9) we also have to place the plane mirror at a comparatively small distance. In this case the adjustment cannot be effected by changing the frequency, as the necessary change would be too large. Instead there is a movable mirror at $D = 90$ m. The mirror being convex, the image to the geodimeter is equivalent to that from the distant mirror.

If K in eq. (8) really was a constant, K could be determined once for all. K however varies somewhat with t_1, t_2 and t_3. The variation hitherto has usually been less than two cms. Nevertheless it must be determined at each individual measurement. It is difficult to make a rapid change from $D = 90$ to $D = 7000$ m. Because of that the distance $D = 90$ is used to calibrate a small loop of light, continuously variable and including the first zero-point where $N = 1$ (eq. 8) and $D = 1$ m or so. In Fig. 3.1 we see the arrangement of the variable loop. The resistances ·0005 in Fig. 3.2 effect an appropriate increase of t_2 and K with it (eq. 8). Thus the zero-point is brought to somewhere near the middle of the variable loop. Unlike the light coming from the distant mirror, that passing through the loop emanates from all parts of the light source. The two paths of light are fully comparable only if the inverse symmetry of the curve in Fig. 3.4 is absolute. Not relying fully on this, we graduate the loop by measurements at $D = 90$ m, and we can do that because a change of K will have the same effect in the loop as in the 90-m path. In all these measurements the value of D is a mean of four determinations as mentioned above. Such a mean did not show any marked change if the light used came from the middle or the one or the other half of the space between the Kerr plates. Thus the situation of the image of L between the Kerr plates need not be exact, and the inversed symmetry of the curve in Fig. 3.4 seems to be very good.

The different parts of the geodimeter will now be described a little more in detail.

DESCRIPTION IN DETAIL

Optics

The source of light L (Fig. 3.1) is the incandescent spiral wire of a Luma projection lamp, T 5, 6 volts, 5 amperes, the dimensions of the spiral being 2 × 2 mm. The position of the holder is adjustable. The single positive lens P_1 is placed half-way between L and the image of L between the Kerr plates. The distance object-image is 22 cm. N_1 and N_2 are ordinary Nicol prisms. The lens P_2 is positive and its focal distance 20 cm. The plano-concave lenses, Nm_1 and Nm_2, focal distance $-$ 12 cm, are silvered on their plane backs and adjustable vertically and horizontally. The mirrors M_1 and M_2 are exactly spherical, diameter 46 cm and focal distance 75 cm. They are cut very accurately ($\pm$ 20'') by Arenco, Ltd. The silvering on the front is applied by vaporization ("Vacuum-Metall"). The spherical aberrations of M and Nm are the same but their signs are opposite. Only yellow light being used, the chromatic aberration is insignificant. The lens P_3 is positive and the focal distance is 2·5 cm. It projects an image some tenths of a millimeter in width in the aperture of the light protecting cope of the phototube. In front of P_3 is placed a green-yellow filter F. It is composed of one green and one yellow ordinary photographic filter.

The plane mirror M_3 is adjustable vertically and horizontally. Its diameter is 33 cm. The reflecting surface has been obtained by aluminium vaporization (AGA). For directing purposes there is a field-glass fitted with a hair cross. The curvature radius of the mirror will probably be larger than 5 kms. The convex mirror Cm of the 90-m measurements (Fig. 3.10) has a curvature radius of 0·5 m. The mirror is movable along a guide over a 50-cm scale S_2.

The variable loop is shown in Fig. 3.1. The small plane mirrors 1, 2, 5, and a convex mirror 4 (radius 6 cm) are fixed. The right-angled mirror 3 is movable on a bar furnished with the 0–70 cm scale S_1. 1 and 5 can be moved aside, when the ordinary measurements are made. 4 being convex, an appropriate quantity of light reaches the photo cathode.

The Kerr cell (1) Kc (Fig. 3.5) has circular electrodes of rustless steel. They are 8 mm in diameter, and the distance between them is 2 mm. The one-piece glass vessel has been made and the electrodes fixed by Freiberger & Andrae. The liquid is nitrobenzene purified by

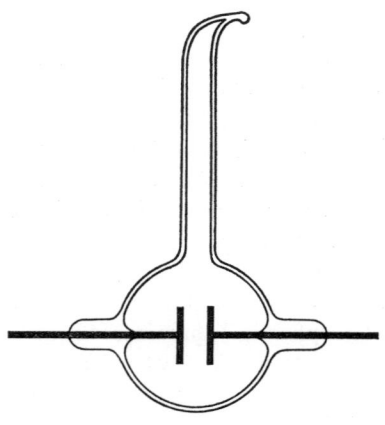

FIG. 3.5.

shaking for three hours with an equal weight of aluminium oxide, threefold distillation *in vacuo* (retort and receiver in one piece), the first and last thirds of the distillate being rejected each time, and finally by high-tension electrolysis for one hour. In order to remove absorbed air the nitrobenzene was shaken in the vacuum before each distillation. Special precautions were made to keep air out of the nitrobenzene during the whole purifying process. After being filled the Kerr cell was sealed by fusing.

With a tension of 5000 volts 50-cycle plus 2000 volts high frequency (8 Mc) the temperature of the cell will rise to some 60° C above that of the surroundings. A small fan reduces this additional temperature to 20° C. After a 300 hours' run there was no observable change in the cell.

The effective wave length of the light used depends mainly on the filter, the transmission of nitrobenzene and the sensitivity of the phototube. The following determination was performed: A slit was placed outside the Kerr cell; with a minimum of deviation, the light was passed through a crown glass prism and a projecting lens. From the resultant spectrum different parts were selected by means of a narrow slit. After passing the filter, F, the selected light was absorbed by the photocell, and the photo-current was measured. With 5 volts

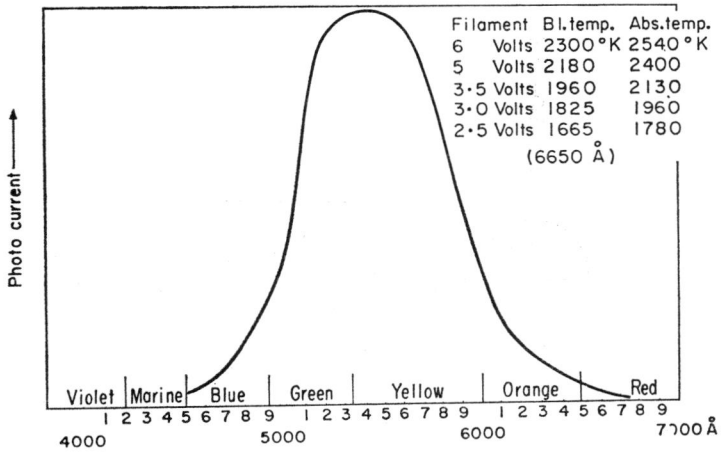

Fig. 3.6.

on the lamp, L, normal running tensions in the Kerr cell, the sensitivity curve shown in Fig. 3.6 was obtained. The curve is corrected with regard to the variation in dispersion with the wave length. The colour was judged by the naked eye. As the successive movements of the slit over the spectrum were always made in equal stages, the adapted and adjusted colour-scale cannot on the whole be displaced by more than 50 Å. When we consider that the index of the refraction of light indicates the change in the velocity in vacuum as compared to that in air, and that the mentioned index includes a term B/λ^2 (λ = wave length) we get from the curve an effective index of refraction corresponding to the wave length 5440 Å at 5 volts on the lamp. Knowing the ordinary temperature of 2540° K at 6 volts (from the manufacturer) we get the black temperature at this and other voltages from the usual tables. Thus the lamp voltage was usually 2·8–3·5 volts during the measurements. According to a table by Pirani this means 1900°–2130° K or 1770°–1960° K black temp. By Wien's emission formula the curve shown in Fig. 3.6 can be transformed to the temperature of 1770° K black temperature corresponding to 2·8 volts on the filament. The effective wave length then has changed to 5570 Å. In view of the blue absorption of the air and of

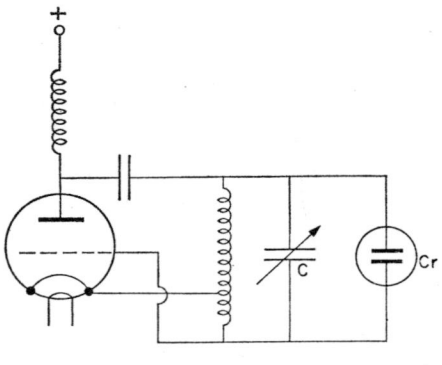

Fig. 3.7.

the fact that this absorption was proportionally larger when higher filament tensions than 2·8 volts were used a refraction index corresponding to 5600 Å was used throughout.

The crystal oscillator with amplifier

The circuit of the crystal oscillator is shown in Fig. 3.7. By the rotating condenser C the frequency can be varied. Cr is the crystal. Fig. 3.8 shows the frequency as a function of the capacity of C. The crystal itself is of an older type held between plane brass electrodes. The protecting box of aluminium contains a contact thermometer and a heating wire. The whole set thus forms a thermostat. The frequency continuously increases by itself during running conditions. The increase seems to be of the expected magnitude. Every measuring night the frequency was checked by wireless transmission to the check station of The Telegraph Service.

The capacity of C having large values, the crystal cools down owing to looser coupling. The consequence is a fall in frequency. Therefore C was kept within reasonable limits. The first hour after the start the increase in the frequency is considerable, and thus the measurements during this period are uncertain.

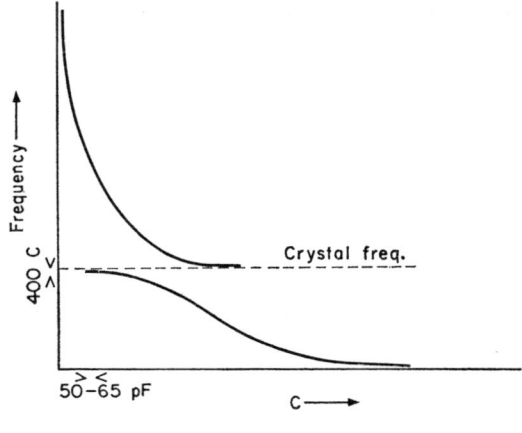

FIG. 3.8.

In Table I (p. 135) are shown all checks of the frequency for the different values on the scale of C. A working curve (Fig. 3.9) of the frequency depending on the C scale has been plotted from Table I, after a parallel reduction of all values to 8332230 Hz ("base frequency") at C = 4·7. No objection to such a reduction can be detected from the table nor is it to be expected within the limited frequency interval of 400 cycles, where the measurements were made. The capacity to the circuit is thereby changed from 50 to 65 pF.

After the quartz oscillator there is the H.F.-amplifier HA (Fig. 3.2) in three steps, the two last of which have push-pull arranged double valves. As is to be seen in Fig. 3.2 the outgoing wires from HA thus come directly from the choked anodes of the power valve, the H.F. power of which is some 40 watts.

The balance circuit

The 70 M Ω resistance of the photo-anode is at the same time the grid resistance of a low-frequency amplifying valve LA (Fig. 3.2) in front of the specific balance circuit. To compensate for the voltage drop due to the direct component of the photo-current, there is a possibility of varying the tension at the earthed end of the resistance. The two valves 1 and 2 of the balance circuit are steep, Philips EF 50. They are chosen as electrically similar as possible. By variation of the screen-grid tension and appropriate positions of the contacts of

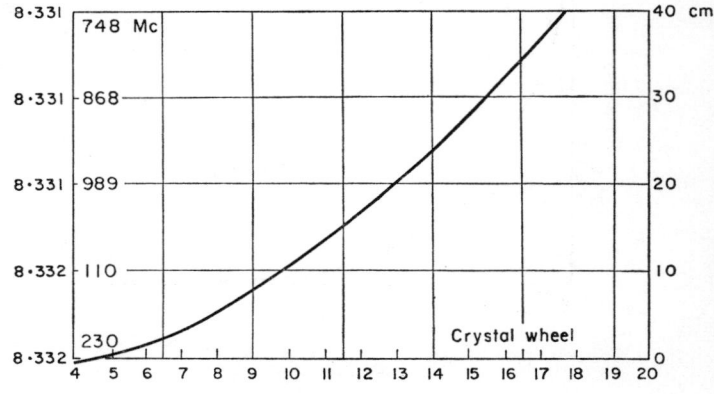

Fig. 3.9.

the potentiometer, 0·01–0·01, 1 and 2 effect equal variations in their anode currents to a certain change of tension on their joined control grids. The suppressor grids, fed by 50-cycle tension from Tr, according to Fig. 3.2, are each directly connected with a diode to ensure normal working tension during the positive 50-cycle half-cycles. The anodes of 1 and 2 are earthed by 10 μF block condensers.

The μA-meter I will give full deflection to 60 μA. The instrument box is filled with paraffin oil to increase the damping effect (10 sec).

The running tensions are smoothed out by neon tubes. Across the secondary coil of the filament transformer there is a potentiometer with a movable earthed tap.

The phototube (RCA 1P21)

The phototube (Fig. 3.2) is enclosed in a lightproof earthed brass cope. The ninth (last) binode is earthed. Proceeding to the cathode the tension is successively increased (by a potentiometer) to -850 volts. The voltage is stabilized by neon tubes. The anode is fed by high-frequency tension, as mentioned above.

Dependence on the running conditions

The distances to the successive zero-points (eq. 8) are increased by 5 mm per one per cent of the increase in the supply voltage (more rapid photo-electrons). These displacements are the most common among those which have to be compensated for by the readings on S_1. The H.F.-amplitude, kept under manual control, has a slight effect on the zero-points too. One per cent increase in amplitude shortens the distances to the zero-points by 0·6 mm.

A change of 500% in the intensity of the received light has no detectable influence on the positions of the zero-points.

Instruments and controls

1. The zero instrument, I, has already been mentioned in "The balance circuit."
2. Voltmeter to the supply voltage.
3. Voltmeter to the lamp L.
4. Wheel to the rheostat of L.
5. Wheel of a turnable copper disk in the tuning coil of the Kerr cell to compensate the resonance change caused by the capacity decrease of the warmed cell. The dielectrical constant of the nitrobenzene decreases.
6. Wheel of a rheostat in the screen circuit of the H.F. power valve, whereby the H.F.-amplitude is kept constant when fluctuations in the supply tension are considerable.
7. M.A.-meter of a small crystal receiver R (Fig. 3.2), showing the H.F.-amplitude and resonance of the Kerr cell circuit.
8. Frequency wheel of the crystal oscillator.
9. M.A.-meter in the anode circuit of the photocurrent amplifier LA, showing the magnitude of the direct current component of the photo-current, including dark current and correct bias tension of LA.

10. Zero instrument in series with I for rough adjustments.
11. Movable taps from the anode resistances of the two balance valves and leading to the zero instruments.
12. Variable screen-grid balance of these valves.
13. Variable earth point of the balance circuit filament current.
14. Variable bias tension of LA.
15. Phase shifter SwI of the photocell H.F.-feed.
16. Phase shifter SwII of the 50-cycle feed of the suppressor grids of the balance valves.
17. Movable mirror 3 (Fig. 3.1) with scale for reading off the position of the first zero-point.
18. Thermometer for reading off the temperature in the crystal oscillator box.
19. Thermometer for the air temperature along the path of light out to the plane mirror.
20. Barometer for the air pressure along the same path.

Readings off

Measuring against the distant mirror, we read off the frequency wheel after each change in the phase shifters, the deflection on instrument I being zero. Zero deflection is equal to that when the plane mirror is screened off. This deflection of course by adjusting the balance circuit, is to be kept close to that caused by zero current through instrument I. After each group of four readings of the frequency wheel, corresponding to the four possible combinations of the positions of the phase shifters, analogous readings S_1 are to be made of the position of the movable mirror at zero deflection of the instrument I. Measuring against the convex mirror at 90 m will be described below.

Among the remaining instruments that for the H.F.-amplitude is to be kept under constant supervision. The wheels (11), (12) and (13), as above, are adjusted once for all. The remaining wheels are usually adjusted before each measurement series. Air pressure and temperature are read off once an hour. These observations are necessary because the velocity is dependent on the atmospheric conditions. This will be dealt with in the next section.

INFLUENCE OF THE ATMOSPHERE

Refractive index of dry air

With the present measurements, the changes in the refractive index of the air with changes in temperature, pressure and humidity play an important role. To transform the obtained value of the velocity to corresponding value for vacuum we also have to know the real magnitude of the refractive index.

The increase in the velocity of light on transition from air to vacuum is supposed to be proportional to the refractive index. This is applicable to the wave velocity. If the refractive index of the air is expressed by

$$\mu = A + \frac{B}{\lambda^2} + \frac{C}{\lambda^4} \tag{11}$$

where A, B, C, are constants and λ the wave length, we get, after taking the present case with *group velocity* (3, 9, 14) into account, a corresponding "refractive index for group velocity" suited to the computations:

$$\mu_g = A + \frac{3B}{\lambda^2} + \frac{5C}{\lambda^4} \tag{12}$$

For dry air at 0° C, 760 mm Hg, Barrell and Sears (2) give the following expression of the wave refraction index:

$$\mu_0 = 1 + \left(2876 \cdot 4 + \frac{16 \cdot 288}{\lambda^2} + \frac{0 \cdot 136}{\lambda^4}\right) \cdot 10^{-7} \tag{13}$$

with λ in thousandths of a millimeter.

After transforming to group velocity, we get, with $\lambda = 0 \cdot 0005600$ mm or 5600 Å:

$$\mu_B = 1 \cdot 00030388 \tag{14}$$

Using the formula of Perard (*Trav. Bur. Int. Poids Mes.* 1934) we get:

$$\mu_P = 1 \cdot 00030382 \tag{15}$$

According to Köster and Lampe (*Phys. Zeit.* 1934):

$$\mu_0 = 1 + \left(2877 \cdot 57 + \frac{15 \cdot 843}{\lambda^2} + \frac{0 \cdot 193}{\lambda^4}\right) \cdot 10^{-7} \tag{16}$$

we get

$$\mu^K = 1 \cdot 00030390 \tag{17}$$

Bureau of Standards' value (1918) used in 1948 (14) is obviously too low. The above figures agree with the mean of 40 values after 1857 (2) and with recent values obtained by the determination of the yard and meter in wave lengths (2). Thus for 5600 Å the following value is used:

$$\mu_g = 1\cdot0003039 \pm 0\cdot0000002 \tag{18}$$

The variations of μ_g with λ, t, p, e

From eq. (12) we get the variation with the wave length as

$$\frac{d\mu_g}{d\lambda} = -0\cdot00000055 \text{ per } 100 \text{ Å} \tag{19}$$

If the temperature is $t°$ C, the pressure p mm Hg and the humidity e mm Hg, the refractive index of air, according to Kohlrausch, will be:

$$\mu_L = 1 + \frac{\mu_0 - 1}{1 + \alpha t} \cdot \frac{p}{760} - \frac{0\cdot00000055\,e}{1 + \alpha t} \tag{20}$$

where $\alpha = 0\cdot00367$ ($\simeq 1/273$).

After replacing μ_0 by μ_g we get the variations of μ_g with temperature, pressure and humidity at 760 mm Hg and

$$0°\text{ C:} \quad \frac{d\mu_g}{dt} = -1\cdot11 \times 10^{-6} \tag{21}$$

$$+10°\text{ C:} \quad \frac{d\mu_g}{dt} = -1\cdot04 \times 10^{-6} \tag{22}$$

$$0°\text{ C:} \quad \frac{d\mu_g}{dp} = +0\cdot398 \times 10^{-6} \tag{23}$$

$$+10°\text{ C:} \quad \frac{d\mu_g}{dp} = +0\cdot384 \times 10^{-6} \tag{24}$$

$$0°\text{ C:} \quad \frac{d\mu_g}{de} = -0\cdot055 \times 10^{-6} \tag{25}$$

$$10°\text{ C:} \quad \frac{d\mu_g}{de} = -0\cdot053 \times 10^{-6} \tag{26}$$

The variation of all these quantities with variations in pressure may be neglected here as being less than 10^{-9} pro mm Hg.

Determination of the atmospheric conditions

To determine the air pressure, a Hg-barometer was placed on a level with the light path. The barometer was checked by the Meteorological Service to be correct to within 0·05 mm.

The temperature was read off by means of thermometers at both ends of the measured distance. They were placed 0 and 2 m above the light path and 4 and 3 m above the ground, which at both ends had an inclination of 1 : 10 down to the fields, situated 15 m beneath the light path. The thermometers were graduated in tenths of degrees C and checked to be correct to within ± 0·03° C.

As regards the temperature gradient above the ground, Johnson and Heywood (20) have values for every month of the year and every hour of the day, obtained above ground similar to that in the present case. Allowing for a slight bend of the isotherms along the ground, the one for the light path has been supposed to pass 4 m above the site of the thermometer place. The temperature gradient was estimated with Johnson and Heywood's curves, to be some + 0·05° C per meter, thus giving an average correction of + 0·2° C in the temperatures read off.

The humidity was estimated to be 70% in the spring nights and 85% in the autumn nights.

MEASUREMENTS ON THE FIELD

Adjustments

The image of the lamp filament is arranged to fall between the Kerr plates. The optical axis of the mirrors and lenses are arranged to coincide with each other. When the back Nicol has been temporarily replaced by a small inclined mirror, the image of some distant object, the top of a tree or the like, is to be seen in the mirror between the Kerr plates. By moving the mirror Nm_1 (Fig. 3.1), the optical system is focused until the image remains clear of parallax. The searcher field glass of the unmoved geodimeter is then adjusted until it is directed towards the same object. By use of the searcher and with unchanged transmitting optical system we then can direct the geodimeter, i.e. the beam of light towards the distant mirror. At this end of the distance the cooperator systematically sweeps the perpendicular of the mirror plane over the geodimeter region by small turnings of the adjusting screws of the mirror. After a few minutes the reflected light is to be seen as a glimpse at the geodimeter end and with the help of agreed light signals the mirror soon will be adjusted towards the receiving optical system. By means of a small inclined auxiliary mirror in front of the filter F the receiving optical system is adjusted so that the focused beam falls upon the aperture of the lens P_3 (Fig. 3.1). In the case of a strongly reflected light or a specific signal lamp, the beam will be seen as a projected spot. This facilitates the adjustments.

The position of the first zero-point

To determine the position of the first zero-point in relation to the geodimeter index, measurements were performed over a distance of

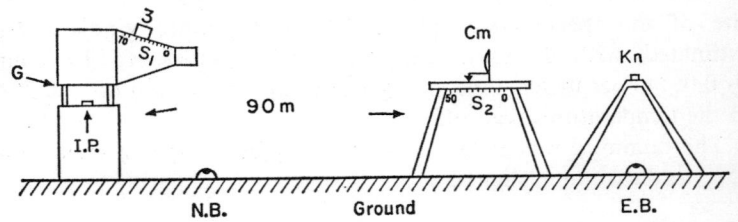

Fig. 3.10.

90 m. In Fig. 3.10 we see the arrangement at the northern end of the Enköping base-line.

G is the reference index, placed in the middle of the bottom back edge of the geodimeter. I.P. is the index knob of the concrete instrument pillar. 3 is the movable mirror of the variable loop having the scale S_1 0–70 cm. N.B. is the northern ground knob of the 7 km base-line. Cm is the small movable convex mirror on its foot with the scale S_2 0–50 cm. E.B. is the ground knob of the small "evaluating base" at a distance of 90 m from I.P. Kn is an index knob, plumbed from E.B. G, I.P., N.B., Cm and E.B. are all in the vertical plane of the 7 km base-line or its extension. I.P., S_2 and Kn are on a common straight line.

The specific measurements were made in the following way:

On the scale S_1 was read off the position of mirror 3 at zero-deflection on the instrument L. Four readings were obtained corresponding to each of the four possible position combinations of the shifters SwI and SwII. The mean of the four readings was the value taken. With the image of the lamp filament projected towards the mirror Cm, a corresponding series of readings of its positions were made on the scale S_2, and the mean was taken. Knowing the distance Cm–G reduced by $10/4 \cdot \lambda$ (eq. 9), we get the distance between the first zero-point and G for a certain position of mirror 3 at I-zero reading. All readings S_1 and S_2 are assembled chronologically in Table II (p. 137). S_1 are in italics. The following table gives the distance values for the 90-m determination in May:

I.P.–E.B. horizontally by invar tapes and strings	+ 91·0885 m
Corr. of strings + 0·40 mm per 24 m	+ 0·0012
Corr. of tapes 1·495 mm per 4 m	+ 0·0075
Sum	+ 91·0972 ± 0·0005 m

An altitude difference of 11·709 m causes the used distance to be

I.P.–E.B.	+ 91·8466 m
E.B.–Kn horizontally (by theodolite)	+ 0·0009
Kn–S_2 (zero line of S_2)	− 1·6470
Cm index–Cm surface	+ 0·1220
I.P.–G	+ 0·1320
Corr. of optics focused to ∞	+ 0·0120
S_1–S_2 as a mean of 13 readings (Table II)	− 0·0526
$10/4 \cdot \lambda$ at 8° C, 760 mm Hg, 8332240 Hz	− 89·9232
The distance G–1st zero-point Sum	+ 0·4907 ± 0·0021 m

With this position of the first zero-point we should obtain zero-deflection on the μA-meter, I, at the movable mirror 3 directed on the zero line of S_1.

A direct measurement of the light path in the loop at mirror position 0 also gave 0·49 m. This determination is none too certain, but corroborates the value above and shows that the curve of Fig. 3.4 has good inverse symmetry.

In autumn the corresponding figures were:

I.P.–E.B.	+ 91·8466
E.B.–Kn	+ 0·0040
Kn–S_2	− 1·6512
Cm index–Cm surface (new setting)	+ 0·1235
I.P.–G	+ 0·1285
Corr. from focus	+ 0·0120
S_1–S_2 as a mean of 12 readings (Table II)	− 0·0600
10/4 . λ at 3° C, 765 mm Hg, 8332180 Hz	− 89·9230
The distance G–1st zero-point Sum	+ 0·4804 ± 0·0021 m

The difference of 1 cm as compared with the spring value may very well be due to changed dimensions of the light path within the geodimeter. They were however unaltered during the respective measuring periods. Unfortunately the light path in the loop was not measured again in the autumn.

The position of the 769th zero-point

In the course of the long-distance measurements, the position of the 769th zero-point in relation to the index G is the same as the position of the plane mirror relative to G. That is so because we adjust the frequency until the zero-point and the mirror coincide. The main part of the distance from G to the plane mirror is the fixed distance from I.P. to the southern ground knob of the base-line. The determination of this distance is shown in the following table:

N.B. knob–S.B. knob on the ellipsoid according to the manuscript G 4102 (of the Survey)	+ 6885·0672
Considering the measurements of the Baltic Geodetic Commission	+ 0·0028
Considering safe (spare) markings	− 0·0020
Corr. for the actual mean altitude of 27·4 m.a.s. (G and mirror)	+ 0·0296
Corr. for the diff. in alt. betw. G and mirr. = 1·1 m	+ 0·0002
N.B.–I.P. by invar tapes	+ 13·2441
Corr. of tapes = 1·495 mm per 4 m	+ 0·0049
Used distance I.P.–S.B. knob Sum	+ 6898·3468 ± 0·0030 m

Then we have the adjoining distances between the S.B. knob and the mirror and between I.P. and G.

The determination of the distance between the S.B. knob and the mirror is shown in Fig. 3.11. 8 m backwards in the extension of the base-line a pile was driven into the ground. The pile had an index knob P. At right-angles 17 m to the west there was placed the stand of the mirror, 3·5 m in height and furnished with a table and index knob K. The horizontal distance P–K was 16·92 m. The angle α

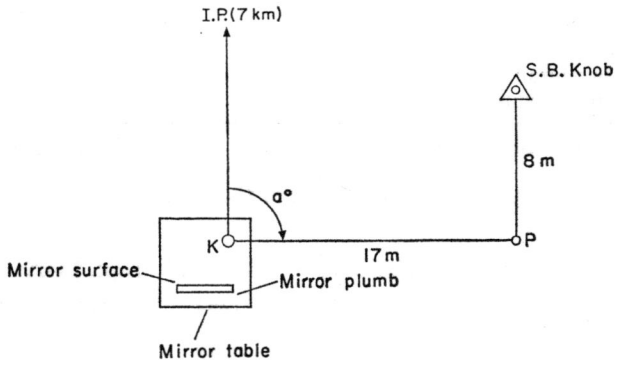

Fig. 3.11.

was measured with a theodolite. The accuracy corresponded to 0·1 mm in the direction of the base-line. Supposing the distance I.P.–P was exactly equal to I.P.–K, α would be $99^c\,92'\,20''$. α as measured was $100^c\,04'\,70''$ in May and $100^c\,05'\,52''$ in September*. The differences from $99^c\,92'\,20''$ give a 3·32 and 3·54 cm respectively shorter distance I.P.–K than I.P.–P.

In a rough calculation in May, the quantity 3·32 cm was used with a wrong sign. The value of the velocity obtained 299 795·6 (this was possibly published somewhere) agreed with that of 1948, which was very misleading.

The distances between the S.B. knob and the mirror are entered in the two following tables, the first one for May, the second for September:

---

* The right angle is here divided into 100 grades: $100^c = 90°$. The grade is further subdivided by hundredths.

S.B. knob–plumbed index	− 0·0019
S.B. index–P by invar tapes	+ 7·9753
Corr. of invar tapes	+ 0·0030
Distance from the angle α (Fig. 3.11)	− 0·0332
K–mirror plumb	+ 0·1648
Plumb–mirror surface	− 0·0225
Total distance S.B. knob–mirror in May	+ 8·0855 ± 0·0005 m
S.B. knob–plumbed index	+ 0·0006
S.B. index–P by steel tape	+ 7·9755
Corr. of steel tape	+ 0·0015
Angle α	− 0·0351
K–mirror plumb	+ 0·1630
Plumb–mirror surface	− 0·0225
Total distance S.B. knob–mirror in Sept.	+ 8·0830 ± 0·0005 m

In the following table are put together the total distances between G and the mirror

Date	May 6–11	May 13–14	Sept. 24–29	Oct. 3
G–I.P.	+ 0·1426	·1245	·1208	·1216
I.P.–S.B. knob (see above)	+ 6898·3468	·3468	·3468	·3468
S.B. knob–mirror (see above)	+ 8·0855	·0855	·0830	·0830
G–mirror	6906·5749	·5568	·5506	·5514 m

Reduced distance from zero-point to zero-point

The adjustments of the position of the 769th zero-point on the plane mirror by means of the freq. wheel are effected for different values of the frequency (increase of fr. p. 9, 23). Furthermore, these adjustments were made under varying atmospheric conditions. To make the measurements comparable we make a reduction to 0° C, 760 mm Hg dry air and the base frequency 8332230 Hz with wheel set at 4·7 (Fig. 3.9). After the reductions, owing to the uncertainty of the determinations, the 769th zero-point will lie at a somewhat varying distance from the plane mirror.

When calculating such a distance we start with the readings on the frequency wheel. All these long-distance readings are assembled chronologically in Table III (p. 138). The groups of four (or eight, etc.) readings are taken directly from the frequency wheel. Among these groups we see the S_1-readings (in italics) of the movable loop mirror 3. The hours of start and conclusion for each night are inserted too. The thermometer and barometer readings are assembled in Table IV (p. 140).

From the curve of Fig. 3.9 we get the distance of the 769th zero-point from the plane mirror with the frequency wheel set at 4·7. As an example we take the first readings in Table III. The figure 11·6 corresponds to a difference in frequency of 185 Hz as compared with that obtained with the wheel set at 4·7. Since the total distance from G to the plane mirror is 6906 m, with 185 Hz we get a displacement of the 769th zero-point of

$$-\frac{185 \times 6906}{8\,332\,000} = -0.153 \text{ m or } -15.3 \text{ cm.}$$

The minus sign indicates that the point lies on the hither side of the mirror. In this way the curve in Fig. 3.9 also can be graduated in such cm-distances. This is done in the scale along the right margin. The three following readings on the wheel were made with the three other combinations of the shifters SwI and SwII (Fig. 3.2). As a mean of the first group of four readings we get the distance 13·9 cm. However, we need the actual position of the first zero-point as determined by the position of the movable mirror 3 (Figs. 3.1, 3.10). Here, too, we have four readings (S_1) in a group. Usually we take the mean of the groups before and after the group of wheel readings. As the before going group is missed in the start we only take the succeeding one and get an S_1 of 16·5 cm. After a reduction with this quantity the whole series of zero-points will start (= 1st point) at a distance of 49·07 cm from G (from p. 121). Inserting temperature, pressure and 70% humidity (Table IV) in the formulae (p. 118) we get the corrections − 6·1 cm, − 2·6 cm and − 0·2 cm.

Thus we get

Distance as from the curve in Fig. 3.9	− 13·9
Movable mirror (S_1)	+ 16·5
Temp. corr.	− 6·1
Corr. of temp. corr. (p. 118)	− 0·1
Pressure corr.	− 2·6
Humidity corr.	− 0·2
Sum	− 6·4 cm

The mean of all determinations on the 6th May is -4.4 ± 0.63 cm. On an average the frequency can be considered to be 5 Hz above the values of the curve in Fig. 3.9. This means a further corr. of $+0.4$ cm. As result on the first night we get the 769th zero-point at 0° C, 760 mm Hg dry air and 8332230 Hz to be -4.0 cm on this side of the plane mirror if the first zero-point had caused an S_1-reading $=0$. Thus the total reduced distance between the 1st and the 769th zero-point will be:

1st zero-point–G (p. 121)	− 0·4907
G–mirror (p. 124)	+ 6906·5749
Mirror–769th zero-point (above)	− 0·0400
1st zero-point to 769th zero-point	+ 6906·0442 m

CALCULATION OF THE VELOCITY OF LIGHT

The above value of 6906·0442 m is the distance D_N in eq. (9), N being 768. From the eq. (10) with $n = 8\,332\,230$ Hz, we find the velocity in dry air at 0° C, 670 mm Hg to be

$$c_a = 299\,701.8 \text{ km/sec.}$$

The effective wave length of the light used being 5600 Å, we get, according to eq. (18) p. 118, the relation to the velocity *in vacuo* to be

$$\mu_g = 1.0003039$$

Thus the final result from the first night is

$$c = 299\,792.9 \text{ km/sec.}$$

In the same way we can compute the rest of the table on the other page. To begin with a computation has been made of all reduced distances from the 769th zero-point to the plane mirror. These distances have furthermore been corrected to hold for the first G–mirror position May 6–11 having the distance from first zero-point to plane mirror equal to $6906·5749 - 0·4907 = 6906·0842$ m. From the mean error for each night has been obtained a certain weight, valid for the value obtained for that night, the starting-point being the inverse square of the night's mean error. In order that the measurements for each separate night might have their full individual effect, the greatest weights were reduced to about half and the lowest weights somewhat increased.

In the table on p. 128 the obtained cm-distances are assembled, with their weights and the number of determinations. Thus to get the zero-distances D_N (eq. 9) we have to reduce 6906·0842 by the different cm-distances, and the velocity is found from eqs. (10) and (18). To illustrate the step-wise increase of the frequency with working time the means of the frequency corrections used each night are also inserted in the table. There is also a column giving the running voltage of the lamp. Owing to variable visibility, the intensity of the return light was the same all the time.

During the whole spring visibility was good. In the autumn, however, only on the two last days. No systematic influence of the haze was observed. Probably a deepening red was compensated for by the higher temperature of the lamp.

From the table we get

Mean of spring values	299 793·04 ± 0·19 km/sec.
Mean of autumn values	299 793·10 ± 0·20 km/sec.
Total mean	299 793·1 ± 0·14 km/sec.

$$\left(\pm \sqrt{\frac{[p\,v\,v]}{[p] \cdot (13 - 1)}} \right)$$

The limits are obtained from the deviations from the mean. To get the real limits we have to consider the uncertainty of several external conditions.

The error in the colour estimation has been set at 50 Å. According to eq. (19) this causes a relative error of $\pm 2·8 \times 10^{-7}$.

The mean error for the temperature will be less than $\pm 0·2°$ C, that of the pressure less than $\pm 0·2$ mm and that of the humidity less than ± 1 mm. Thus from eqs. 21–26 we get an atmospheric error of $\pm 2·5 \times 10^{-7}$.

The Enköping base-line was remeasured by the Baltic Geodetic Commission. The resulting mean error of 4×10^{-7} will be increased to 5×10^{-7} by including errors in the small actual added distances.

Regarding the frequency, the mean error will not be larger than 2 Hz or $\pm 2·4 \times 10^{-7}$.

Date in 1949	Lamp voltage	Fr. corr. cm	Number obs.	Weight 1 obs. = 1·0	Dist. cm	Velocity 299 790 +	Temp. C°	Pressure mm Hg
May 6	2·8	−0·4	4	2·0	4·0	2·87	+ 8	750
8	2·8	−0·4	6	12·0	2·4	3·56	+ 1	755
9	2·8	−0·6	1	1·0	4·9	2·50	+ 3	762
10	2·8	−0·7	7	5·0	4·5	2·63	+ 9	765
11	2·8	−2·5	5	9·0	4·8	2·52	+ 8	769
13	2·8	−2·5	5	3·0	2·5	3·52	+11	765
14	2·8	−2·4	6	2·0	3·8	2·93	+ 9	763
Sept. 24	5·0	−3·3	2	0·5	4·1	2·84	+11	768
26	5·0	−4·5	5	2·0	0·7	4·27	+15	764
27	5·0	−4·5	5	5·5	4·4	2·68	+12	763
28	6·0	−4·5	2	0·3	1·3	4·05	+11	764
29	3·5	−5·2	5	7·5	4·1	2·84	+14	758
Oct. 3	3·5	−5·3	5	8·2	3·0	3·32	+ 9	749

Finally the position of the first zero-point contributes with an error of $\pm 3.0 \times 10^{-7}$.

Thus totally we get the errors

769th zero-point	$\pm 4.5 \times 10^{-7}$
1st zero-point	$\pm 3.0 \times 10^{-7}$
Refr. index	$\pm 2.0 \times 10^{-7}$
Colour	$\pm 2.8 \times 10^{-7}$
Atmosphere	$\pm 2.5 \times 10^{-7}$
Base-line	$\pm 5.0 \times 10^{-7}$
Frequency	$\pm 2.4 \times 10^{-7}$

$$\sqrt{[vv]} = \pm 8.80 \times 10^{-7}$$

As a final result of the determinations on the 7 kms base-line we arrive at a velocity of

$$c = 299\ 793.1 \pm 0.26 \text{ km/sec.}$$

The limits include all known contributory errors.

Values from other base-lines

1. As the base-line above was bisected (angle 198° 84′ 86″) there were two further distances which could be used for determinations. The longer one, with the instrument still at N.B., and having 573 zero-points, was composed of the following distances.

1st zero-point–G	− 0·4804	
G–I.P.	+ 0·1227	
I.P.–M.B. (along light path)	+ 5147·7036	± 0·010
M.B.–mirror	− 3·7576	
Mirror–573rd zero-point (reduced)	− 0·0045	
1st to 573rd zero-point	5143·5838	± 0·013

This gives the value of c as

$$c = 299\ 794.0 \pm 0.75 \text{ km/sec.}$$

2. Using the shortest base-line, with the geodimeter at M.B., the distance components were:

1st zero-point	− 0·4804	
G–M.B.	+ 4·1550	
M.B.–S.B.	+ 1750·9060	± 0·0070
S.B.–mirror	+ 7·9175	
Mirror–197th zero-point (reduced)	− 0·0214	
1st to 197th zero-point	1762·4767	± 0·0090 m

The velocity being

$$c = 299\ 792·3 \pm 1·5 \text{ km/sec.}$$

3. With practical experience of the complete geodimeter, the 1948 determinations (14) can now be corrected.

Then, in the absence of the 90-m distance, the first zero-point can be determined with tolerable accuracy from a direct measurement of the light path in the loop, as we now know. Thus from the data of the records it is evident that the first zero-point as determined at Linköping must be put forward by 7 cm.

Instead of the temperature corrections of $0·93 \times 10^{-6}$ employed (Kohlrausch at 20° C), $1·1 \times 10^{-6}$ is more correct.

Regarding the colour, a curve has been obtained with the actual filter and the filament voltage used as shown in Fig. 3.12. A value of μ_g of 1·000304 seems to be correct. The used base frequency of 8332157 Hz (younger crystal), checked a few days (12/10, Table I) after the measurements, is correct.

The reduced distances of 1948 were

1. 1st–1009th zero-point: 9064·377 m
2. 1st–469th zero-point: 4208·430 m

at 5 mm humidity giving

1. $c = 299\ 796·3 \pm 1·5$ km/sec.
2. $c = 299\ 794·1 \pm 3·0$ km/sec.

Mean $c = 299\ 795·8 \pm 1·5$ km/sec.

After reduction of the distances by the above mentioned corrections to a total of respectively 7·8 and 7·4 cm, the corrected values of 1948 are

1. $299\ 794·1 \pm 1·3$ km/sec.
2. $299\ 789·3 \pm 2·5$ km/sec.

Weighted mean: $299\ 793·1 \pm 2·0$ km/sec.

4. Finally we have the distance measurements of 1947.(18) These measurements were performed with the old geodimeter model. The crystal oscillator, however, was transferred in one piece to the new model, so that we can find the frequency used. From Table I it can be seen that the frequency has increased by 150 cycles from 12th October 1948 to 14th October 1949. The running time is estimated at some 150 hours from the measuring records.

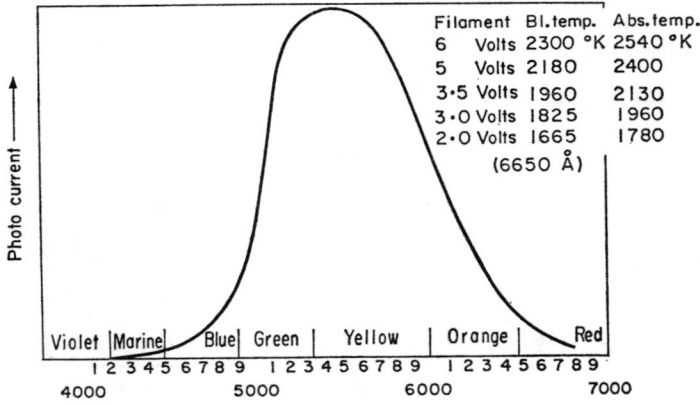

Fig. 3.12.

The running time during the preceding year will be about 10 hours at Lovö, 60 hours at Linköping (from the records) and 50 hours of experimental work at AGA (entries about working hours). To this must be added some 10 hours of demonstration. We get 130 hours as the total or 130 cycles. The frequency will then be 8332027 ± 30 Hz.

A strong support for the correctness of this method of procedure is the fact that the frequency in 1942 as given by the manufacturer (note on the crystal) was 8·3315 Mc. 1942–1947 the crystal was employed for experimental work. A running time of 500 hours during the period mentioned seems to be reasonable.

From the paper (18), p. 108, is seen that the first zero-point in 1947 was based on direct measurement of the light path, and as we now know this is satisfactory. From the same paper, p. 110, we have the "unit" length 17·98494 m at 0° C, 773 mm Hg. If we correct for humidity and pressure and transform to vacuum value we get $\lambda = 35·98088$ m. Then we have used a refractive index of $\mu_g = 1·000302$ due to gold-plated mirrors and photocell 1 P 22 with sensitivity reaching into red. With $n = 8332027$ Hz we get the velocity obtained in 1947 to be

$$c = 299\ 793·9 \pm 2·7 \text{ km/sec.}$$

The point Vårby (p. 110) being of second order, a value from the distance 7734 m should assume the weight practically equal to 0.

Assembling all the measurements in a table, we get

Year	Distance	Weight	Velocity	
1947	11025 ± 0·08 m	0·1	299 793·9 ± 2·7	93·9
1948	9064 ± 0·01	0·6	94·1 ± 1·3	⎱ 93·1
1948	4208 ± 0·01	0·2	89·3 ± 2·5	⎰
1949 May	6906 ± 0·0035	8·0	93·0 ± 0·27	⎫
1949 Sept.	6906 ± 0·0035	8·0	93·1 ± 0·28	⎬ 93·1
1949	5144 ± 0·01	1·5	94·0 ± 0·75	
1949	1762 ± 0·01	0·4	92·3 ± 1·50	⎭

Weighted mean: 93·1 ± 0·25 km/sec.

The error limits of the distances denote the geodetic accuracy.

Final result

The foregoing investigation yields the value of the velocity of light *in vacuo* to be

$$c = 299\ 793·1 \pm 0·25 \text{ km/sec.}$$

No sign of any systematic variation with time can be discovered.

COMPARISON WITH DETERMINATIONS BY OTHERS

MICHELSON (1926) (4, 11) obtained

$$c = 299\ 798 \pm 4 \text{ km/sec.}$$

BIRGE's statistical collation in 1942 (10, 11) gave the result

$$c = 299\ 776 \pm 3 \text{ km/sec.}$$

Essen's determination (1948) (13) by short radio waves in resonance in a short guide gave the value

$$c = 299\ 792 \cdot 2 \pm 4 \cdot 5 \text{ km/sec.}$$

(Essen gives ± 9 as the maximum error.)

Aslakson's value (1949) (15) by radar applied to six geodetically known distances in U.S.A. is

$$c = 299\ 792 \cdot 4 \pm 2 \text{ km/sec.}$$

where the inner mean error of $\pm 1 \cdot 5$ km/sec. here has been increased to ± 2 to allow for possible systematic errors.

DETERMINATIONS OF LENGTHS

As a practical application two uncertain 1st order triangle sides in Norrland were determined.

1. The distance between the islands of Prästgrundet and Storjungfrun near Söderhamn in August:

1st night	20 203·63 m.	One obs.
2nd night	20 203·57 ± 0·04 m.	Six obs.
Mean	20 203·59 ± 0·04 m.	

The coordinates give 20 203·79. However the length as determined from the new Söderhamn base-line via a base net and two triangles is 20 203·25 m. Supposing this value to be of double the weight of that from the coordinates, the geodetically determined length would be 20 203·43 m which is in tolerable agreement with the value determined with light.

2. The distance between the two fjelds Ounistunturi and Sautusvaara near Kiruna in Lappland in September:

1st night	30 921·67 ± 0·17 m.	Two obs. weight 1
2nd night	·52 ± 0·05 m.	Ten obs. weight 3
3rd night	30 921·37 ± 0·07 m.	Five obs. weight 2
Weighted mean	30 921·50 ± 0·07 m.	

The frequency was checked as usual. Provisionally adjusted coordinates gave the distance 30 921·42 ± 0·60 m.

In the Norrland measurements and with the 1762 m base-line, the geodimeter was supplied with tension from a portable 400 watts benzene motor generator. For the rest the 230 volts net was used.

TABLE I

Checking of the frequency. Cr. wh. = readings on the crystal wheel

Date in 1948–49	Hour	Cr. wh.	Freq.	Date in 1948–49	Hour	Cr. wh.	Freq.
12/10	13·45	—	Start	6/5	20·10	12·5	8331 979
	15·00	0	8332 166	contd.	·15	17·5	1 740
	·05	5	157		·20	15	1 879
	·10	10	037		·25	0	2 210
	·15	13	8331 921		·30	10	2 085
	·20	16	779		·35	5	2 210
	·25	20	543		·40	0	2 218
	·30	23	323		21·55	10	2 103
	·35	26	048				
	·40	30	8330 541	9/5	19·45	—	Start
	·45	33	005		20·55	10	8332 080
	·50	36	8329 171				
	·55	40	8327 258	10/5	19·40	—	Start
	16·00	43	8324 200		20·55	10	8332 091
	·05	46	8316 640		21·55	10	110
	·10	50	8272 265				
27/4	14·00	—	Start	11/5	19·45	—	Start
	15·35	0	8332 209		20·55	10	8332 096
	·40	5	2 203		21·55	10	125
	·45	10	2 075				
	·50	15	1 860	12/5	19·50	—	Start
	·55	20	1 557		20·55	12·5	8332 021
	16·00	30	0 489				
	16·10	0	2 170	13/5	19·45	—	Start
4/5	18·00	—	Start		20·55	12·5	8332 012
	19·30	0	8332 200		21·55	12·5	030
	20·05	0	2 219				
	20·10	5	2 211	14/5	18·45	—	Start
	·15	7·5	2 172		20·15	0	8332 246
	·20	10	2 092		·20	5	2 246
	·25	12·5	1 987		·25	10	2 115
	·30	15	1 870		·30	15	1 890
	·35	17·5	1 733		·35	17·5	1 737
	·40	0	2 195		·40	12·5	1 972
					·45	10·0	2 080
6/5	19·07	—	Start		·50	7·5	2 176
	20·00	0	8332 195		·55	0·0	2 231
	·05	7·5	2 152		21·55	10·0	2 122

TABLE I—*continued*

Date in 1948–49	Hour	Cr. wh.	Freq.	Date in 1948–49	Hour	Cr. wh.	Freq.
27/8	20·30	—	Start	29/9	18·40	—	Start
	21·30	10	8332 093		19·55	10	8332 153
					20·55	10	2 168
29/8	19·30	—	Start		21·55	10	2 173
	21·30	10	2 142				
				3/10	18·40	—	Start
10/9	19·25	—	Start		19·55	10	8332 157
	20·55	10	2 122		20·55	10	2 163
	21·55	10	2 133		21·55	10	2 176
				4/10	19·00	—	Start
	19·20	—	Start		19·55	10	8332 135
12/9	22·00	17·3	8331 813		20·55	10	2 140
					21·55	10	2 170
	19·50	—	Start				
23/9	20·55	10	8332 117	5/10	18·45	—	Start
	21·55	10	2 144		19·55	10	8332 153
					20·55	10	2 164
	18·30	—	Start		21·55	10	2 184
24/9	20·55	10	8332 150				
	21·55	10	2 157	6/10	19·00	—	Start
					19·55	10	8332 141
					20·55	10	2 166
26/9	18·50	—	Start		21·55	10	2 175
	20·55	10	8332 160				
	21·55	10	2 167	14/10	12·00	—	Start
					14·00	10	8332 211
27/9	18·45	—	Start		14·02	0	2 339
	19·55	10	8332 141		14·04	10	2 203
	20·55	10	2 157		·06	5	2 327
	21·55	10	2 164		·08	10	2 207
					·10	15	1 987
					·12	5	2 324
28/9	18·45	—	Start		·14	17·5	1 847
	19·55	10	8332 159		·16	10	2 184
	20·55	10	2 165		·18	20	1 691
	21·55	10	2 157		·20	10	2 184

TABLE II

90-m measurements. Readings on S_1 (in italics) and on S_2 (p. 120) in cm. Each reading in a group of four corresponds to one of the four possible position-combinations of the shifters SwI and SwII

May 15		May 19			Sept. 30			Oct. 7		
18·1	22·7	*19·6*	22·4	*17·5*	*15·3*	42·2	*23·0*	*15·6*	29·3	26·4
21·5	19·9	*23·7*	18·8	*21·3*	*23·0*	37·8	*19·2*	*21·8*	25·9	23·8
9·7	22·9	*11·2*	22·6	*10·2*	*12·3*	7·6	*27·2*	*17·8*	24·3	25·6
13·1	21·5	*14·8*	21·4	*14·4*	*23·4*	4·8	*23·3*	*24·6*	24·4	22·9
25·7	*18·1*	20·8	*19·5*	22·5	31·4	33·9	*15·6*	12·7	*31·1*	*14·5*
23·6	*21·2*	16·9	*21·4*	19·9	32·6	29·8	*21·0*	10·4	*23·4*	*19·0*
20·3	*10·3*	25·3	*9·3*	23·7	17·6	17·5	*15·1*	40·6	*22·6*	*18·8*
17·6	*16·2*	23·5	*14·8*	22·4	11·4	12·9	*20·0*	36·6	*20·5*	*24·2*
20·7	*21·5*	19·7	20·9	*18·1*	24·0	*14·9*		23·7	*16·5*	25·4
21·0	*18·4*	23·0	19·4	*22·5*	21·0	*19·1*		20·6	*19·8*	23·0
8·9	22·0	9·8	21·9	*9·8*	28·0	*13·4*		29·8	*17·7*	27·7
10·9	20·9	*14·2*	21·0	*13·3*	26·0	*22·8*		26·8	*23·5*	24·8
22·9	16·8	22·1	*17·7*		12·3	9·8		*14·9*	30·2	25·8
19·4	20·8	17·0	*22·8*		*21·3*	3·0		*18·6*	24·3	20·8
23·9	10·7	25·3	*10·8*		*15·2*	47·8		*18·3*	24·5	25·8
23·6	*13·4*	22·5	*14·8*		*23·0*	42·2		*25·0*	20·1	24·6
17·7		21·3	22·3		27·5	30·2		21·7	19·3	*15·9*
19·3		23·8	19·8		21·7	25·4		18·5	14·2	*21·7*
10·7		10·8	23·4		26·7	22·1		31·3	30·8	*14·7*
11·6		14·3	21·3		24·0	17·4		26·5	28·9	*19·9*
21·5		21·0	20·7		6·4	*11·6*		28·4	*17·7*	22·3
20·1		18·1	*22·1*		1·0	*19·8*		25·2	*20·5*	21·9
22·6		26·0	*9·7*		46·8	*15·9*		25·2	*14·5*	25·8
20·5		23·2	*14·9*		42·6	*21·9*		22·6	*18·2*	24·9
17·1		*21·3*	21·9		*15·3*	19·6		*14·9*	20·8	21·9
21·7		*21·9*	19·8		*19·9*	17·0		*19·9*	17·7	20·7
10·9		*9·9*	22·8		*15·7*	28·6		*16·7*	31·6	27·5
15·9		*14·9*	21·4		*19·3*	25·2		*24·0*	27·4	24·2
										12·5
										17·7
										15·6
										22·4

TABLE III

7-km measurements. Readings on S_1 (p. 120) in italics and on the crystal wheel. Each reading in a group of four corresponds to one of the four possible position-combinations of the shifters SwI and SwII. The first figures each day are date and hour of start. The last figures are hours of conclusion

6/5	8/5	16.5	10.9	9.8	15.4	*21.0*	10.3	*24.4*
		12.8	14.1	11.3	12.6	*17.5*	*17.2*	*9.8*
21.40	21.45	*11.7*	10.0	*12.0*	13.6	*10.3*	*22.4*	*14.1*
		14.7	12.2	*10.7*	11.3	*13.2*	*9.2*	
11.6	*14.7*	*21.0*	*14.8*	*22.0*	*19.7*	*8.0*	*13.8*	23.30
13.3	*13.0*	*21.4*	*12.1*	*17.6*	*21.2*	*9.8*	*12.1*	
11.2	*11.8*		*19.4*		*11.5*	*14.4*	*8.4*	
7.9	*9.1*	23.00	*18.0*	23.00	*14.7*	*11.0*	*12.1*	
16.4	*10.7*		*11.3*		*14.8*	*19.6*	*8.0*	24/9
12.6	*8.1*		*12.9*		*12.5*	*17.7*	*19.1*	
20.2	*15.6*	9/5	*9.9*	11/5	*14.1*	*8.9*	*21.0*	19.25
16.9	*13.2*		*11.8*		*11.6*	*11.8*	*10.8*	
12.3	*12.5*	22.55	*14.2*	21.50	*20.0*	*8.3*	*13.5*	*21.3*
8.7	*15.3*		*11.4*		*23.1*	*11.1*	*13.3*	*28.6*
11.7	*18.9*	*13.0*	*21.8*	*19.4*	*10.5*	*13.7*	*8.5*	*12.9*
8.7	*21.4*	*12.1*	*20.2*	*24.6*	*14.6*	*11.6*	*12.7*	*18.4*
21.0	*10.7*	*13.8*	*11.1*	*9.5*		*23.1*	*9.3*	*11.5*
22.1	*7.0*	*12.6*	*13.5*	*16.8*	23.00	*20.8*	*19.5*	*5.3*
11.3	*15.7*	*18.2*	*9.8*	*15.5*		*9.5*	*22.5*	*16.0*
13.3	*13.4*	*19.3*	*12.5*	*12.1*		*12.5*	*9.8*	*11.8*
11.7	*12.9*	*10.1*	*13.2*	*13.8*	13/5	*8.8*	*13.5*	*20.3*
9.3	*16.0*	*14.6*	*11.0*	*10.7*		*11.3*	*13.5*	*22.8*
12.2	*19.3*		*21.8*	*17.6*	21.30	*14.3*	*10.4*	*14.7*
8.6	*21.4*	23.20	*17.2*	*22.1*		*11.4*	*13.6*	*17.8*
17.7	*10.1*		*14.5*	*12.7*	*13.7*	*21.7*	*9.7*	*12.3*
19.6	*7.3*		*17.2*	*17.4*	*11.1*	*19.6*	*18.9*	*6.9*
12.2	*15.5*	10/5	−2.3	*15.6*	*21.5*		*20.4*	*18.4*
14.3	*13.6*		*8.3*	*12.0*	*17.5*	23.00	*8.8*	*11.4*
11.7	*14.2*	20.45	*13.7*	*13.9*	*10.0*		*11.0*	*19.3*
7.6	*14.1*		*9.4*	*11.3*	*12.5*		*13.8*	*24.3*
12.7	*17.8*	*16.8*	*18.7*	*18.0*	*8.0*		*10.5*	*13.2*
8.8	*20.7*	*13.0*	*19.8*	*22.8*	*11.9*	14/5	*14.0*	*17.1*
20.4	*10.8*	*22.2*	*12.3*	*12.2*	*13.7*		*8.9*	
23.2	*6.6*	*17.4*	*13.8*	*16.7*	*10.5*	21.35	*19.8*	20.40
9.7	*15.7*	*12.2*	*8.8*	*14.9*	*21.5*		*24.6*	
15.5	*13.6*	*13.6*	*11.4*	*13.5*	*19.2*	*20.2*	*8.3*	
	12.8	*7.7*	*15.3*	*13.5*	*10.4*	*23.4*	*13.0*	26/9
23.00	*14.3*	*11.4*	*11.4*	*11.0*	*12.1*	*9.2*	*13.6*	
	17.6	*14.9*	*20.2*	*20.3*	*9.0*	*14.8*	*9.6*	19.40
	20.8	*12.0*	*16.7*	*21.7*	*11.3*	*12.1*	*13.5*	
	11.7	*21.6*	*12.7*	*10.3*	*12.2*	*6.7*	*9.3*	*25.6*
	5.6	*20.1*	*14.0*	*14.6*	*9.6*	*14.3*	*19.6*	*27.7*

TABLE III—continued

14·0	12·9	*8·7*	17·3	12·9	29/9	13·4	11·2	15·8
16·6	10·2	*18·2*	14·7	8·4		10·5	8·8	12·3
14·2	10·3		*11·5*	14·5	19.50	12·5	13·4	10·8
11·4	4·8	22.30	*19·7*	10·7		9·6	10·7	4·0
8·9	13·2		*14·2*	13·7	*27·2*	13·7	*12·3*	15·4
4·0	10·9		*20·4*	8·1	*29·7*	8·7	*17·1*	10·9
9·5	*20·0*	27/9	7·4	14·7	*4·6*	13·8	*18·0*	*11·1*
5·0	*24·4*		−5·8	11·3	*10·0*	8·9	*25·2*	*18·6*
13·4	*6·3*	19.50	17·7	13·5	13·0	*14·6*	22.20	*16·3*
11·1	*14·8*		15·3	9·8	8·5	*18·0*		*22·3*
22·7	10·7	*15·7*	9·7	*17·7*	15·5	*17·1*		15·3
25·8	9·0	*21·5*	−5·8	*21·2*	10·0	*23·4*	3/10	10·6
12·4	13·2	*11·1*	16·3	*8·7*	12·4	13·1		12·2
15·9	11·3	*21·4*	14·4	*17·2*	9·8	9·4	20.20	4·2
9·5	13·4	10·0	*12·8*		12·4	13·0	*10·9*	15·5
6·6	11·2	4·0	*19·5*	22.20	10·3	9·4	*14·4*	11·3
13·6	9·9	15·7	*16·8*		*13·4*	13·2	*17·8*	11·3
10·7	8·0	12·1	*25·2*		*19·8*	9·7	*23·4*	7·1
9·7	*18·5*	8·6	14·2	28/9	*17·6*	14·3	14·4	14·4
6·2	*24·4*	4·0	9·4		*23·0*	10·5	4·4	10·8
13·3	*10·3*	16·9	13·7	19.50	13·4	*10·4*	13·5	12·4
11·0	*15·3*	13·4	8·5		10·5	*15·8*	3·9	7·1
22·1	14·8	*11·4*	*11·0*	15·7	12·6	*21·2*	12·5	14·9
27·2	12·6	*16·9*	*22·2*	9·3	9·7	*26·5*	−5·8	9·7
13·3	9·1	*20·3*	*9·6*	8·3	13·2	12·7	15·1	12·7
17·5	7·5	*26·4*	*20·0*	4·0	10·5	8·8	10·3	8·4
18·2	6·4	9·3	13·9	16·4	12·3	13·7	11·1	*11·4*
25·3	10·6	−5·8	10·5	15·7	9·2	11·3	5·8	*16·5*
10·4	11·9	15·9	13·6	8·3	*10·9*	12·0	14·6	*16·0*
14·0	13·9	13·1	10·2	(−5·8)	*15·7*	8·6	10·4	*22·0*
10·7	*18·2*	7·2	14·1		*18·7*	13·4	10·5	
7·0	*25·0*	−5·8	10·0	20.20	*23·4*	11·2	4·0	22.30

TABLE IV

Atmospheric conditions

1949		Temperature		Pressure[1] mm Hg	Temp. of barom.
Date	Hour	Geod.	Mirror		
6/5	21·02	+ 9·2° C	+ 8·6° C	749·3	+ 10·8° C
	21·35	7·9	8·6		
	22·20	7·7	7·6		
	23·05	7·6	7·4	749·7	9·5
8/5	21·25	1·6	2·6	754·9	4·1
	22·15	1·2	1·9		
	23·00	0·4	1·6	755·0	3·3
9/5	21·00	3·9		761·4	5·6
	22·50		2·2		
	23·15		1·6		
10/5	20·57	9·1	8·9	764·6	10·1
	21·45	9·0	9·1		
	22·20	8·6	8·6		
	22·50	7·7	7·8	765·0	9·6
11/5	21·40	8·2	7·8	678·8	9·5
	22·05	7·8	7·8		
	22·40	7·6	6·8		
	23·10	7·1	6·8	768·8	8·7
13/5	21·20	11·4	11·4	764·6	14·1
	22·00	11·4	11·7		
	22·35	11·0	10·3		
	22·55	10·8	10·6	764·6	13·3
14/5	21·25	9·8	9·7	762·3	12·8
	21·50	8·9	8·3		
	22·30	8·7	8·1		
	23·20	8·3	7·5	762·1	10·6
24/9	19·08		12·6		
	19·20	11·1		767·7	12·9
	20·10	10·6	12·1	767·9	12·7
	20·55	11·9	11·9	767·8	12·9
	22·00	11·2	10·9	768·0	12·8

[1] The pressure figures are unreduced readings. Lat = 59°32′, 26 m. a. s. Brass scale, correct at 0° C.

TABLE IV—*continued*

1949		Temperature		Pressure[1] mm Hg	Temp. of barom.
Date	Hour	Geod.	Mirror		
26/9	19·20	+ 15·6° C	+ 15·6° C	764·2	+ 15·9° C
	19·55	15·2	15·3	764·2	15·9
	20·50	14·8	14·7	764·1	16·2
	22·00	14·2	14·7	764·1	?
	22·15	14·0			
	22·30		14·7		
27/9	19·20	13·3	13·7	763·4	14·6
	20·15	12·2	12·5		
	21·00	11·7	12·0	763·1	12·7
	22·00	11·7	11·9	763·0	12·2
	22·00	11·2	11·9		
28/9	19·05	10·8		764·0	12·6
	20·00	10·4	11·8	764·1	12·2
	20·35		10·1		
29/9	19·35	13·8	13·0	758·9	14·6
	20·00	13·6	12·9	758·6	14·5
	21·00	14·2	14·2	758·3	14·7
	22·00	14·1	13·9	757·8	14·6
	22·25	14·3	14·3	757·6	15·1
3/10	19·45	9·9	9·9	747·3	10·6
	21·00	8·9	9·1	749·3	9·8
	22·30	8·1	7·8	751·7	9·5

[1] The pressure figures are unreduced readings. Lat = 59°32′, 26 m. a. s. Brass scale, correct at 0° C.

REFERENCES

1. WHITE, *Rev. Sc. Inst.*, Jan. 1935. (Kerr cell.)
2. BARRELL and SEARS, *Phil. Trans. Roy. Soc. London*, **1**, 1939.
3. JENKINS and WHITE, *Fund. of Phys. Opt.* (Group velocity.)
4. MICHELSON, *Astr. Journ.*, **65**, 1 (1927). (Velocity of light.)
5. MITTELSTAEDT, *Ann. d. Phys.*, **2**, 285 (1929).
6. MICHELSON, PEASE and PEARSON, *Astr. J.* **82**, 26 (1935).
7. ANDERSON, *Rev. Sc. Inst.*, **8**, 239 (1937).
8. HÜTTEL, *Ann. d. Phys.*, **37**, 365 (1940).
9. ANDERSON, *Jour. Opt. Soc.*, **31**, 185 (1941).
10. BIRGE, *Rep. on Progr. in Phys.*, **VIII** (1943).
11. DORSEY, *Trans. Am. Phil. Soc.*, **34**, 1, (1944).
12. WARNER, *Austr. Jour. of Sc.*, Dec. 1947.
13. ESSEN and GORDON-SMITH, *Proc. Roy. Soc.*, **194**, 348 (1948).
14. BERGSTRAND, *Arkiv f. M.A.F.*, **36** A, 20 (1949).
15. ASLAKSON, *Trans. Am. Geoph. Un.*, **30**, 475, (1949).
16. HOUSTOUN, *Nature*, **164**, 1004 (1949).
17. BERGSTRAND, *Arkiv f. M.A.F.*, **29** A, 30 (1943). (Meas. of dist.)
18. ——, *l'Act Com. Geod. Baltique*, Helsingfors 1948.
19. HART, *Bull. Geod.*, nr 10, 1948.
20. JOHNSON and HEYWOOD, Met. Off. Geoph. Mem. IX, 1938. (Temp. gradient.)

Index

Acoustic diffraction grating 16
Air, refractive index 4, 62, 117ff, 126
Anderson 17, 82
Arago 6
Aslakson 18

Baseline instability 12
Baseline measurement 58ff, 121ff, 133
Bennett 37
Bergstrand 25ff, 32, 101ff
Birge, reviews 17, 34, 82, 132
Bradley 3

Cavity resonator 20, 21, 22, 84ff, 91ff
Cleland 33
Cornu 5
Crystal oscillator 112

Diffraction grating, acoustic 16
Dorsey 34, 82, 83

Essen 23, 133
Essen and Gordon-Smith 18, 19ff, 80ff
Evacuated tube 11, 50ff

Fizeau 5
Florman 33
Forbes 5
Foucault 6
Frequency measurement 20, 89ff
Froome 32, 36, 37

Geodimeter 25, 32, 38, 101ff
Group velocity 4, 19
Guenther 34
Gutton 16

Hale 9
Houstoun 16
Hüttel 17

Jastram 33

Karolus 17
Kerr cell 16, 25, 26, 30, 102ff
Knutsen 37
Krypton wavelength standard 36

Laser 36
Luckey 33

Mackenzie 32
McKinley 16
Metre, standard 36
Michelson, early work 6ff
Michelson, Pease and Pearson 9ff, 43ff, 82
Microwave oscillator 90
Microwave receiver 89
Mittelstaedt 17

Newcomb 6

Pearson 9ff, 43ff
Pease 9ff, 43ff
Perrotin 5
Phase velocity 4, 19

Quality factor, Q 22, 23, 85, 90ff, 99

Rank   34
Römer   3
Rosa   34, 82, 83
Rotating mirror method   5, 13, 45ff, 53ff
Rotating wheel method   5

Saksena   34
Schöldström   32
Shearer   34
Silsbee   34

Velocity of light, results   74, 75, 98, 127ff

Waveguide   19, 20
Weil   33
Wheatstone   5
Wiggins   34

Young   5

Selected Readings in Physics

There is a tendency nowadays for undergraduates to learn their physics completely from textbooks, not becoming acquainted with the original literature and therefore not realising how the subject grew and developed. The purpose of the series in which this volume is published is to present a set of reasonably priced books which give, for a particular subject, reprints of those papers which record the development of important new ideas.

KINETIC THEORY
Part A: The Nature of Gases and of Heat

Stephen G. Brush
A.B.(Harvard); D.Phil.(Oxon.)
Lawrence Radiation Laboratory, Livermore, California

Intended as an aid to the teaching of physics from the historical viewpoint, this book includes reprints of selected works on the theory of gases and the nature of heat by Boyle, Newton, Daniel Bernoulli, George Gregory, Robert Mayer, Joule, Helmholtz, Clausius, and Maxwell.

Although for the most part these works expound modern views on the conservation of energy and the kinetic theory of gases, an attempt is made to illustrate the older views that have been replaced by these modern theories. Thus the Boyle–Newton repulsive theory of gases is presented, as well as the mean-free-path explanation of gas viscosity, the Maxwell velocity distribution law, and the virial theorem. The papers selected have been written in a fairly clear and simple style, and should therefore be readily appreciated by the modern reader; they include qualitative descriptions of phenomena and statements of philosophical positions, as well as mathematical derivations. The Introduction provides the proper historical context for the understanding of the reprints.

Selected Readings in Physics

MEN OF PHYSICS: L. D. LANDAU I
D. ter Haar

L. D. Landau can justly be described as one of the world's most eminent physicists. He was awarded the Nobel Prize in Physics and although this was for his work on the theory of condensed media, and especially for his work on the theory of liquid helium, the Prize might equally deservedly have been awarded for other work. Plasma physics, high-energy physics, quantum mechanics, and the theory of magnetism are all topics to which Professor Landau has contributed greatly. Although comprehensively covered in the Collected Papers of L. D. Landau recently published*, it is felt desirable that a selection of some twenty of his most important papers should be made available in the form of two inexpensive publications.

This first volume contains eight papers: two on the theory of helium II, two on the theory of Fermi liquids, two on superconductivity, one on electron diamagnetism and one on ferro-magnetism. The second volume contains twelve papers: one on the theory of phase transitions, one on stellar energy, one on the statistical model of nuclei, one on the multiple production of particles in cosmic rays, one on the uncertainty principle in relativistic quantum mechanics, two on the quantum theory of collisions, two on plasma physics, and three on field theory.

* *The Collected Papers of L. D. Landau.* Edited by D. ter Haar. Pergamon Press.